THATCHERISM AND BRITISH POLITICS

Dennis Kavanagh is Head of the Department of Politics at the University of Nottingham, and co-author of the last five Nuffield ESRC-sponsored studies of British general elections. He has written more than a dozen books, including *British Politics: Continuities and Change* (OUP, 1985) which was described by the *Times Higher Education Supplement* as 'a leading candidate for the basic text in any university course on British politics', and *The Thatcher Effect: A Decade of Change* (OUP, 1989), edited with Anthony Seldon.

By the same author:

Constituency Electioneering in Britain
Political Culture
The British General Election of February 1974 (with David Butler)
The British General Election of October 1974 (with David Butler)
The British General Election of 1979 (with David Butler)
British Politics Today (with W. Jones)
New Trends in British Politics (editor with Richard Rose)
The Politics of the Labour Party (editor)
Political Science and Political Behaviour
The British General Election of 1983 (with David Butler)
Comparative Government and Politics: Essays in Honour of S. E. Finer (editor with Gillian Peele)
British Politics: Continuities and Change
The British General Election of 1987 (with David Butler)
Consensus Politics (with Peter Morris)
The Thatcher Effect (edited with Anthony Seldon)

THATCHERISM
AND
BRITISH POLITICS

The End of Consensus?

Second Edition

DENNIS KAVANAGH

OXFORD UNIVERSITY PRESS
1990

Oxford University Press, Walton Street, Oxford OX2 6DP

Oxford New York Toronto
Delhi Bombay Calcutta Madras Karachi
Petaling Jaya Singapore Hong Kong Tokyo
Nairobi Dar es Salaam Cape Town
Melbourne Auckland

and associated companies in
Berlin Ibadan

Oxford is a trade mark of Oxford University Press

Published in the United States
by Oxford University Press, New York

First Edition published 1987 in hardback and Oxford paperbacks
Paperback reprinted 1987, 1988 (twice), 1989
Second edition published 1990 in hardback and Oxford Paperbacks

British Library Cataloguing in Publication Data

Kavanagh, Dennis
Thatcherism and British politics: the end of consensus—
2nd ed.
1. Great Britain. Politics
I. Title
320.941

ISBN 0-19-827756-3
ISBN 0-19-827755-5 (pbk)

Library of Congress Cataloging-in-Publication Data

Kavanagh, Dennis.
Thatcherism and British politics: the end of consensus?/Dennis Kavanagh.—2nd ed.
Includes bibliographical references.
1. Thatcher, Margaret. 2. Great Britain.—Politics and government—1945- I. Title.
DA591.T47K38 1990 941.085'8'092—dc20 89-28844

ISBN 0-19-827756-3
ISBN 0-19-827755-5 (pbk.)

Printed in Great Britain by
Biddles Ltd, Guildford & King's Lynn

To M.

PREFACE TO THE SECOND EDITION

I AM delighted to have this opportunity to write a second edition of *Thatcherism and British Politics*. The publishers will also be relieved, after four reprints of the first edition. At the time when I started this book in May 1984 some friends doubted that there was such a subject as consensus politics and other sceptics thought that Thatcherism would be short-lived. In fact Thatcherism was only one aspect of the book. As Jim Bulpitt pointed out in a review in *Political Studies*, the book's main concern was to explore the interplay between political ideas and practice in twentieth-century British politics, and to show how the political agenda had changed in the 1970s. By 1989 Mrs Thatcher is still in office and her government's impact on the political scene demands serious study.

This edition leaves intact the chapters dealing with the historical background (2), new right ideas and think tanks (3 and 4) and the changes in the circumstances and political parties (5, 6, and 7) which helped to break up the post-war political settlement. Much that is relevant to the later chapters of the book has happened since I finished writing the first edition in April 1986. I have therefore completely re-written chapters 1, 8, 9, and 10 in the light of these developments. For helpful comments on the new edition I am grateful to Jane Kavanagh and Peter Morris.

D. K.

30 June 1989

PREFACE TO THE FIRST EDITION

This book is not a contribution to the 'wet' versus 'dry' debate in the Conservative party. If asked, my reply to Mrs Thatcher's famous question 'Is he one of us?' would, on the whole, be 'no'. Nor is it intended to be a detailed analysis of the record of the 1979 government. Excellent studies of this have already been written by Peter Riddell in *The Thatcher Government* (1983) and Peter Jackson and his colleagues in *Implementing Government Policy Initiatives: The Thatcher Administration 1979–1983* (1985). It does not aspire to compete with the numerous biographies of Margaret Thatcher (too many of which border on the reverential).

This book accepts that there was something called 'the political consensus' that was accepted by most leaders in the parties, prevailed through the 1950s and 1960s, and gradually broke down in the 1970s. It is difficult to assign with precision responsibility for the breakdown of the consensus, though Mrs Thatcher wanted to break it and could claim some credit for doing so. The book studies the nature of the consensus and the factors that led to its emergence, explores the causes of its decline, and analyses the subsequent impact of Thatcherism on British politics. It is not a book about Thatcherism or Mrs Thatcher but tries to place both in the broader background of post-war British politics.

The idea for the book was occasioned by the subject of my inaugural lecture at Nottingham University in May 1984. Such lectures often deal with the professor's views on the state of his discipline, or his part of it, and with his hopes and plans for its future development at his institution. For various reasons the timing did not seem appropriate for such a lecture. I therefore chose to address a topic that was of contemporary concern but could also be studied in a broader context. The lecture 'Whatever Happened to Consensus Politics?' was later published in *Political Studies*, 33 (December 1985). I have drawn on parts of the article in Chapters 2 and 5. Some part of Chapter 9 on Mrs Thatcher's style of prime ministerial leadership is based on my paper *Margaret Thatcher: A Study in Prime Ministerial Style*, published in January 1986 as a professional paper in the University of Strathclyde Studies in Public Policy (No. 151).

I am indebted to a number of organizations and individuals for help. The Nuffield Foundation provided a small grant to cover travel and interviewing expenses. Between November 1984 and June 1985 I was the fortunate recipient of a Visiting Fellowship at the Hoover Institution on War, Revolution and Peace. This enabled me to study and write in congenial surroundings at Stanford University in California. I took advantage of the intellectual stimulus provided by permanent and visiting fellows and am especially grateful to Dr Jerry Dorfman and Dr Peter Duignan both for arranging the visit and for good companionship. I am also grateful to Gallup for granting permission to use survey material, David Robertson for use of Fig. 2.1, to S. Brittan and P. Lilley for permission to reprint Table 2.3, and to P. Whiteley and I. Gordon for permission to use Table 6.3. I am grateful to those politicians and officers of groups who agreed to grant interviews.

Many friends have been kind enough to comment on earlier versions of some chapters of this book. For such help I am grateful to Vernon Bogdanor, Nicholas Bosanquet, Monica Kavanagh, John McClelland, Michael Moran, David Regan, and John Wood. My colleague Peter Morris has been particularly generous in his willingness to read and challenge what I have written. In spite of their sterling efforts I must claim sole responsibility for whatever faults remain.

D. K.

30 April 1986

CONTENTS

TABLES

FIGURE

1

INTRODUCTION

BRITAIN has long been regarded as a model stable democracy. How many would willingly exchange the British experience in the twentieth century for such other models as Japan, Germany, Italy, or France? In the years after 1945 it achieved an increase in living standards and a rate of near full employment that would have amazed people living in the inter-war years. In terms of economic and social progress the post-war years were, by British standards, a success story. However, in the mid-1970s there was a more pessimistic mood, in which Britain was seen as a warning to other western states. In a farewell interview in 1976, President Ford warned: 'It would be tragic for this country if we went down the same path and ended up with the same problems that Great Britain has.'[1] The interaction of high inflation, low economic growth, trade union power, and weak government all amounted, critics claimed, to 'the future that does not work'.[2] Marxists wrote of a crisis of the capitalist state, and both they and Conservatives referred to a crisis of social democracy. But, on the whole, right-wing calls for the simultaneous restoration of government authority and reduction in the size of the public sector were heard more clearly.

The Thatcher Experiment

Since 1979 the so-called 'Thatcher experiment' in Britain has attracted attention, partly because Mrs Thatcher's Conservative governments have broken with so many features of the post-war consensus and partly because there are many conflicting assessments of the governments' record. Leaders of some right-wing parties in Western Europe have been impressed by the Thatcher government's re-election in 1983 and 1987 and have canvassed the political and economic benefits of a regime of markets, monetarism, and authoritative government. By contrast, critics in Britain (including some Conservatives) have damned the record, particularly the rising unemployment and loss of manufacturing capacity; or have claimed alternatively that the governments have not carried out their programme (particularly in maintaining tight monetary control or more drastically cutting public spending and personal taxes); or have asserted, finally, that the policies have had little impact. The political left has presented a separate indictment of economic failure, divisive policies, authoritarianism, and incompetence.

The so-called Thatcher experiment has certainly made British politics more interesting and it is not surprising that there have been several scholarly as well as polemical studies. The literature on her is now immense. Perhaps only Churchill and Lloyd George, both at times of national crises, have seized the imagination to the same degree. But few books about them were written when they were in office. She had been the subject of at least a dozen biographies by late 1988. Another dozen studies were available for the tenth anniversary of her premiership in May 1989. Lord Callaghan and Sir Alec Douglas Home have each had one biographer and Mr Heath has had but three. Publishers eagerly snap up books on the lady and all her works, and usually ensure that her name (and photograph) dominate the title-page. The attention to her personality as a means of understanding public policy is interesting in itself. It was Suetonius, in his *Life of the Caesars*, who warned that a concentration on personality was a symptom of a declining political order.

There are at least three solid accounts of the record of the Thatcher governments,[3] sets of critical essays published by Marxists and neo-Marxists,[4] and a blistering attack on the government's economic policy by a former Cabinet minister.[5] Most of the biographies border on hagiography. (A notable exception is the study by Hugo Young.[6]) Also relevant are some of the outpourings of the 'New Right', which are discussed below in chapters 3 and 4. Many recent books about developments in party politics, public policy, local government, the economy, the civil service, and so on, also emphasize the role of the Thatcher governments, in breaking the post-war mould of British politics.

No doubt, thirty or forty years hence, we will have a much fuller picture of the events of these years. Historians will have access to a greater volume of government papers which are at present private, as well as to memoirs of the participants. The passage of time provides its own balance (and distortions) as well as opportunity for the longer-term consequences to work themselves out. But this does not excuse us from making an attempt to understand ourselves in our own time, not least before particular myths are established. In 1945 R. B. McCallum, co-author of the first 1945 Nuffield study of a British general election, was inspired by his memory of the myths surrounding the 1918 general election (for example, 'Hang the Kaiser') to make an authoritative study of the 1945 election. He felt that such a study should be written close to the event to prevent the implantation of such myths. His belief was that the election 'must be photographed in flight, studied and analysed'.[7]

As a co-author (with David Butler) of five successive studies of British general elections I have gained some experience in trying to explain and interpret general election outcomes. The intense but episodic study of election campaigns and interviews with élite participants does of course have some disadvantages. Many of the major events in post-war British politics have not actually been associated with general elections. One thinks of the devaluation of the pound in 1949, the Suez invasion in 1956, the reappraisal of Conservative policy in 1961 which led to the application to join the European Community and the introduction of economic planning, the second devaluation of the pound in 1967, the introduction of statutory incomes controls in 1972, and the IMF's rescue package for the British economy in 1976. In themselves elections are rarely turning points in the political affairs of a nation.

In 1983, however, it was clear, to this writer at least, that a number of important changes had taken place in British politics. Compared to 1974 the questions of trade union power and runaway inflation were virtually absent. The apparent ungovernability of Britain and the government's lack of authority also seemed issues that belonged to a bygone age. In addition, there had occurred the shift in style and policy of the Conservative party and the continuation of the long-term electoral decline of the Labour party. Such established ground rules for electoral politics as, for example, that no British government could win an election with one million, let alone over three million, unemployed, or that the Conservatives had to be led from the centre to be electorally competitive, or that some form of corporatist arrangements were necessary for industrial regeneration and curbing inflation, also lay in ruins. These changes can also be connected with the now rather inconclusive debates about Britain's long-standing relative economic decline and the erosion of its manufacturing base. The 1983 election did not cause these changes in the political agenda but rather confirmed the change of direction which had already occurred.[8]

It is trite to say that we have been living through a period of social, economic, and political change. In *The Strange Death of Liberal England* (1935) George Dangerfield tackled the question of how it was that the Liberal party, securely in office in 1914 and the creator of so many welfare and political reforms, was, by 1919, already declining to the status of a minor party. In contrast to suggestions that it was the war which ruined the Liberal party, Dangerfield observed how, in the years 1910–13, several groups—suffragettes, trade unions, Conservative peers, and Irish nationalists—rejected liberal values and turned to militant tactics. At the apparent height of its power Liberalism was already in retreat and the party doomed to a speedy decline. Have we been witnessing a similar decline—this time of what might be termed Social Democratic Britain?

The expression 'social democracy' is employed here as an umbrella term to embrace four features: a *mixed economy*, which provides for state intervention and management of a market economy for social ends; deference to certain *interests*, the trade unions and working class; their representation by *political parties*, particularly the Labour party; and the pursuit of certain *policies*, particularly of full employment, mixed economy, and welfare state. The policies were also often termed Keynesian or Butskellite (an amalgam of R. A. Butler, the Conservative Chancellor of the Exchequer

1951–5, and Hugh Gaitskell, his Labour predecessor and shadow).

There have been many obituaries for British Social Democracy, notably in the 1930s after the collapse of the second Labour government and again in the burgeoning Marxist literature since the late 1960s.[9] Yet for a time in the mid-1970s Labour appeared to be the natural party of government in Britain. This was so, not in the sense of its having more electoral support than other parties (already the signs were of Labour's long-term decline), but in the sense that its collectivist values were more suited to the prevailing political formulae of functional representation and group politics. Edward Heath's reversals of policy after 1970 only reinforced this view.

In coming to power in 1979 Mrs Thatcher made no secret of her determination to break with what *The Times* has called 'the old clubbable consensus'. She was the product of a change in British politics and, in turn, she gave voice and direction to the new policies. One always has to take care in analysing the rhetoric of politicians, whose speeches are made on different occasions for different purposes and to different audiences. A reader of Mrs Thatcher's speeches, however, does not need to be so careful. She is what she says she is. A content analysis of her speeches in the 1979 general election campaign reveals an emphasis on two themes. The negative one is 'socialism', the positive one is 'freedom', or 'a free society' or 'freedom under the law'.[10] The same themes were present in later speeches. In her rhetoric she scorns consensus politics and proclaims her determination to scrap many aspects of it. Her 'wet' critics in the Conservative party, particularly Edward Heath, Francis Pym, and Sir Ian Gilmour, regularly decry her abandonment of consensus, both in style and in policy substance. Bishops of the Church of England, former Conservative Prime Ministers, many of 'the great and the good', and, more predictably, trade union spokesmen and leaders of the political opposition, have all spoken about the need for consensus and conciliation. This was particularly so during the twelve months of the coal miners' bitter strike in 1984–5. The 'conciliators' sensed that there was something new and regrettably confrontational about the style of government and that this was to be deplored. The Liberal leader David Steel told the House of Commons on 31 January 1985: 'I believe in the politics of persuasion, while she believes in confrontation. We believe in building a consensus.' In speaking thus, Mr Steel was surely in the mainstream of political practice in Britain as it existed between 1940 and 1979.

Consensus Politics

Consensus is not an ideal term because it can be used in so many different ways. My main interest is to identify a set of policies or values which, to a large extent, were shared by all post-war governments until 1979. Trevor Smith has fairly objected to the use of the term consensus which he associates with agreement by deliberation or conscious bipartisanship.[11] He thinks a more accurate term in this context is policy coincidence. However, it seems that consensus is a firmly established and wide-ranging concept. In referring to political consensus students have usually had in mind one or all of the following.

1. A high level of agreement across the political parties and governing élites about the substance of public policy. There have been many areas—defence, foreign affairs, regional policy, Northern Ireland, and so on—where this has often and clearly been the case. But for most of the post-war period the term has also referred to broad agreement and continuity between governments on the mixed economy and welfare state.

2. A high level of agreement between the political parties and governing élites about the nature of the regime or about the rules of the political game. Institutional change in Britain has been incremental and over the past sixty years there has been little popular support for proponents of comprehensive, far-reaching constitutional innovation.

3. The political style by which policy differences are resolved, namely a process of compromise and bargaining and a search for policies which are acceptable to the major interests. Disagreements have rarely been pushed to breaking point and the legitimacy of the government rarely called into question.

The élites have managed to make timely concessions to new interests, though there were near-misses in 1914 in Ulster and 1926 over the General Strike. Consensus has also referred to the tendency for a new government to accept its predecessor's legislation—even when, in opposition, it had derided it. The classic case of this is the record of Conservative governments between 1951 and 1964, which substantially accepted the 1945 Labour government's initiatives in the fields of public ownership, welfare, and the retreat from Empire.

A political consensus, and associated concepts like hegemony, folkways, political culture, rules of the game, and so on, also represents a mobilization of bias. It favours certain interests and directs attention to certain issues and procedures while neglecting others. Here is what social scientists now call 'a second face of power', one that shapes the political agenda. One has only to think of the negative reception given, until recently, to many of the ideas of the free-market wing of the Conservative party and, today, to the ideas of Labour's left wing. In economic management the dominance of Keynesian ideas has for a long time resulted in there being only a perfunctory hearing for proponents of monetarism.

Mrs Thatcher has clearly wanted to break with many old policies. In a prepared speech in the 1979 general election she compared herself to the Old Testament Prophets, who did not say 'Brothers I want a consensus', and she proclaimed the importance of conviction and principle in politics—as if these were incompatible with consensus. In 1981 in Australia, she replied to criticisms from Mr Heath that she was abandoning consensus politics: 'For me, consensus seems to be the process of abandoning all beliefs, principles, values and policies.' The term has long been a 'hooray' word, along with 'moderate', 'centrist', and 'reasonable' and contrasted with 'ideological' or 'extreme' or 'dogmatic'. Some part of the explanation for her hostile reaction to it, is that she realizes (correctly) that it is often a code-word for criticism of her own political style and policies. The consensus has also protected interests, policies, values and a style—many of which she has wanted to abandon. Here is a political leader who deeply believes that the pursuit of consensus in the past has too often involved the appeasement of pressure groups and avoidance of tough decisions and that such avoidance has brought the country to a sorry pass. Her negative assessment of consensus politics and her disavowal, in contrast to most other leaders, of seeking such a goal, follows from this belief.

As a style of policy-making, consensus had already attracted critics in the 1960s and 1970s. The emphasis on consultation with interests, incremental change, and the demonstrated ability of important pressure groups to veto change amounted, some critics alleged, to pluralistic stagnation or political inertia.[12] The style too often was one of make do and mend, sacrificing long-term objectives for the short term. British government places a high premium on surface unity; witness the muzzling of potential Cabinet dissenters by the doctrine of collective responsibility or the emphasis on consultation and agreement in the top ranks of the civil service. What emerges may be a 'directionless consensus'. Consider two critical comments about policy-making in the 1960s:

> . . . public policy is simply equated with finding the least controversial course between the conflicting interests of vociferous private groups. It is not a doctrine of government: it is a doctrine of subordination.[13]

> Maintaining harmony between the disparate group interests is valued much more highly than advancing towards the goals of any particular group or party.[14]

There were two broad reactions to this awareness of interdependence, diffusion of power and weak government.[15] One was to call for a 'strong' political leadership, which would stand up to pressure groups and lower expectations of what government can do. Both Mr Wilson and Mr Heath, for example, tried to tackle trade union power by legislation and appeals to public opinion. The former retreated from the proposals outlined in *In Place of Strife* in 1969, when he found that he could not carry his party or Cabinet. Mr Heath's almost de Gaulle-like use of a general election in 1974 failed to defend his statutory incomes policy against the strike by the National Union of Mineworkers. The alternative was to call for a coalition-monger, for political leaders skilled in power sharing and conciliating different interests. Later (chapter 9), I refer to these approaches as appealing, respectively, for a political mobilizer and for a political conciliator. In 1979 or 1989 there was little doubt which role Mrs Thatcher envisaged for herself and her government.

Thatcherism

There are problems in analysing Thatcherism, not least because we have to separate Margaret Thatcher's personal beliefs and goals from those of her administration, and distinguish these from what is attributed to them by malicious opponents as well as fervent supporters. The term Thatcherism is used in three different contexts. The first refers to Mrs Thatcher's no-nonsense style of leadership and hostility to the premium placed on gaining agreement by consensus. As a leader she is remarkable for regarding politics as a suitable arena for the expression of personal beliefs. In contrast to Harold Macmillan who thought that people should turn to bishops for moral leadership Mrs Thatcher sees the expression of fundamental beliefs as part of the leader's task.

A second usage refers to a set of policies designed to produce a strong state and a government strong enough to resist the 'selfish' claims of pressure groups *via* law and order, traditional moral values, a stable currency, and a free economy (*via* cuts in state spending and taxes, reducing state intervention, and privatization).

The third use of the term refers to the international reaction against high inflation, trade union militancy, and unease about ungovernability in the mid-1970s. The economic recession and slow economic growth undermined popular support for the welfare consensus in a number of other states. The Thatcher governments' policies of tax cuts, privatization, 'prudent' finance, squeezing state expenditure, and cutting loss-making activities has had echoes in other western states.

Such terms as 'Thatcherism', 'monetarism', and the 'New Right' are also often interchanged as if they are the same thing. A monetarist believes that excessive increases in the supply of money (that is, above the increase in production in the economy) cause inflation, a case of too much money chasing too few goods. Too often in today's political rhetoric, however, the term refers not to an approach to economic analysis but to a right-wing Conservative, who questions Keynesian policies and is a free-marketeer on other policies. But there are varieties of monetarism, and monetarism is analytically quite separate from, and has no necessary links with, a market economy, high unemployment, lower public spending, balanced budgets, and so on. Similarly the term 'New Right' is too often used to lump together various social and economic doctrines and policies and political personalities. It is important, however, to separate those libertarians who favour a reduction in the role of the state in both social and economic areas from those who are concerned with the restoration of the authority of the state and hostile to many aspects of 'permissiveness'.

A determinant of a leader's political style is his or her political beliefs. Few doubt that Mrs Thatcher has a coherent set of political ideas and that these guide her behaviour. She often relates them to her early experiences rather than to abstract theory, occasionally invoking early memories of her father, her school days, and even her headmistress (see below, chapter 9). From her secure family background come the values of self-reliance, hard work, thrift, the family, and a belief in just deserts and not looking to the 'nanny state'. These notions have been subsequently reinforced by her reading of some of the economic writings of Milton Friedman, particularly on incomes policies and his advocacy of monetary control and the free market, and the philosophical works of Friedrich von Hayek. She has also turned for intellectual support to the ideas of Sir Keith Joseph and the competitive-market ideas of think-tanks like the Institute of Economic Affairs and the Centre for Policy Studies. But she is not simply a borrower of ideas. A remarkable statement of her beliefs is contained in a speech she gave to the Conservative Political Centre in 1968. In it she argued that British government had become too interventionist and that it was

> totally unacceptable that government should decide the increases of wages and salaries. Conservatives should aim to reduce the range of government decision-making and restore greater individual choice

> . . . What we need now is a far greater degree of personal responsibility and decision, far more independence from the government, and a comparable reduction in the role of government.

In the same speech she said:

> There are dangers in consensus; it could be an attempt to satisfy people holding no particular views about anything. It seems more important to have a philosophy and policy which because they are good appeal to a sufficient majority.

These beliefs come out in her party conference speeches and, more explicitly, in interviews. To a lesser extent they are reflected in the actions of her government. The main elements in her belief system can be stated in terms of certain propositions:

(1) the limited capacity of government to do lasting good, but its great capacity to do harm;
(2) the importance of individual responsibility and the existence of right and wrong (she is one of the few Conservative Cabinet ministers to vote regularly for the restoration of the death penalty);
(3) the state should be strong enough to perform its 'primary' tasks of ensuring adequate defence and law and order;
(4) people should solve their own problems (or help their families and neighbours solve theirs), rather than turn to the government;
(5) increasing public expenditure, without economic growth, involves more taxation and less choice for individuals;
(6) the market is the best means of promoting economic growth and free choice and of safeguarding personal liberty;
(7) more expenditure on one service usually means less on another, unless one resorts to borrowing or inflation. Each service is paid for by 'hard-pressed' taxpayers, many of whom may be poorer than the beneficiaries of particular programmes;
(8) government intervention may be counter-productive in terms of slowing down society's ability to adapt in a

changing world; 'correct' programmes (in terms of the above) are more useful than expressions of sympathy for the weak, unemployed, and sick.

Circumstances have helped some of these ideas to gain acceptance. By the mid-1970s, rampant inflation, industrial disruption, and the growing influence of trade unions in Britain made a number of people willing to try an alternative approach. Keynesian techniques of economic management did not appear to have an answer to 'slumpflation'. The simultaneous slowdown of economic growth and accelerating inflation made it difficult to finance many state programmes, particularly in social welfare. Belief in the efficacy of the main planks of the post-war consensus—high public spending, the mixed economy, and incomes policies—was declining in both major parties. The Left gained ground in Labour, particularly after the party's election defeats in 1970 and 1979, and the Right gained in the Conservative party with the election defeats in 1974 and the exit of Mr Heath. In other western countries also, economic policies of cash limits, curbing the growth of public expenditure, controlling the money supply, and extending the role of the market gained greater acceptance. In 1976, partly in response to IMF pressure, the Labour government made cuts in planned public spending, introduced monetary targets, and continued with incomes policy.

Thatcherism is a matter both of style and policies. As a political style—emanating from Mrs Thatcher herself—it vigorously challenges many established beliefs and interests, boldly expresses personal and often right-wing views, and does not compromise on many deeply held political principles (see chapter 9 for further discussion). As a set of policies it rests on four main planks. First, there was the determination to reduce the increase in money supply so that inflation would be squeezed out of the system. This obviously involved the abandonment of formal incomes policies and 'deals' between government, employers, and trade unions as part of an attack on inflation. If the two sides of industry settled for a rate of increase in wages higher than the growth in productivity, they ran the risk of losing markets and pricing themselves out of jobs. A second plank was the reduction of the public sector and encouragement of a free market-orientated economy. This involved policies of privatization (or sale to the market) of state-owned industries and services, removing 'stifling' regulations on business, and encouraging the sale of council houses. The government would not bail out or subsidize loss-making industries indefinitely. Lower public spending would facilitate tax cuts and these, in turn, would encourage economic growth and the creation of new firms. The third plank was for the government to free the labour market through reforms (pre-strike ballots, periodic and secret ballot elections of union leaders, restriction of secondary picketing, and removal of some immunities which unions had long enjoyed under common law) and encourage responsible trade union practices. Control of the money supply had to work in tandem with the creation of a more effective free-market economy. As Sir Keith Joseph once claimed in opposition, 'monetarism is not enough'. The above measures would, it was promised, eventually create the enterprise society and economy. According to one of the most articulate exponents of 'the new Conservatism', Nigel Lawson, 'Our chosen course does represent a distinct and self-conscious break from the predominantly social democratic assumptions that have hitherto underlain policy in post-war Britain.'[16]

A final plank linked all three of the above policies: the restoration of the authority of government. This involved both strengthening the nation's military defence and forces of law and order, and resisting damaging claims of interest groups. There would be significant increases in resources for the armed forces and police. Interest groups would be curbed by limiting the responsibilities of government in maintaining full employment and the firm control of public spending. In all, the four policies married the values of a strong state and a free economy.[17]

The ideas, particularly those about relations between individuals and state, amounted to a forceful assault on the Butskellite consensus, so much of which, it was claimed, failed to address the major problems of the country and even added to them. It had led people to have expectations of the state (for example, the 'welfare state mentality') which could not be satisfied *in principle* for there is no limit to the money that might be spent on services, just as there is no end to the demands which could be made for services. But many of the services were inadequate anyway, and only the virtual state monopoly of such services as health care and education prevented people from voting with their feet. Critics also claimed that the state was seen less as an enforcer of rules and more as a provider of benefits, an agency which cushioned people from the sometimes harsh—but ultimately beneficial—disciplines of work and the market. But if the state stepped in to ease the costs of adversity then how could people learn from their mistakes? How might one evoke what Keynes once called 'the ancient vein of Puritanism'?

What followed from this analysis of what had gone wrong was the need for a restriction of the role of the state *vis-à-vis* society and economy. The government would not, could not, provide full employment in a free society and it would not subsidize market 'failures'. It would control the money supply, to produce 'honest money', and maintain law and order. In the spirit of John Locke government would no longer provide 'positive' goods, but a 'negative' one, namely the sense of security upon which the attainment of all other, privately chosen goods depends. In the new society there would be a high degree of personal liberty (income tax reduced, liberty under the law, more scope for private enterprise, and so on). Government would be authoritative, doing what it could and should do, for example directly curbing inflation but not unemployment. It would not, however, be a *permissive* society; the counterpart of a strong if limited government is that parents, teachers, managers, and other authority figures are also respected. In many ways this is an old liberal view of the state, but it is also radical in that the analysis goes to the root of problems. In contrast to the old consensual language it sharpens rather than blurs distinctions between state, society, and individual.

There have been different assessments of the impact of the Thatcher government to date. William Keegan is fairly representative of many Keynesian critics of the economic record. In his *Britain Without Oil* he argues that although the government has used the benefits of North Sea oil to build up an overseas portfolio, it has also presided over a catastrophic reduction of manufacturing output, which resulted in the first peacetime manufacturing deficit in history in 1983 (£2.5 billion). Between the first quarter of 1980 and second quarter of 1981 Britain's gross domestic product (GDP) fell by 3.7 per cent, industrial production by 9 per cent, and unemployment nearly doubled to 2.5 million. What occurred after 1979 was 'the compression of perhaps twenty further years of relative decline into five, and all in spite of our possession of North Sea Oil, which was supposed to give the country a big opportunity'.[18]

Such negative verdicts should not cause surprise. Disappointment has been a fact of life for British (and most other) governments. Each post-war government, even the more successful ones of 1945–50 and 1955–9, has been overwhelmed by criticisms soon after leaving office. Labour governments were accused by many on the left of betraying socialism, and Conservative leaders by the right-wing of betraying 'true' Conservatism. Much disappointment has centred on the record of the economy. The Thatcher government has encountered similar charges from Conservative 'wets' who claim that the government is destroying the heritage of One Nation Conservatism. I leave aside for the moment whether these claims have merit.

The most thorough study of the 1979–87 governments by Peter Riddell argues that they fell far short of most of their central objectives.[19] By the 1983 general election a number of commentators took the view that the government had lost its radical cutting edge. 'Radical Thatcherism is dead', wrote the *Guardian*'s Peter Jenkins, noting that the election pledges which Mrs Thatcher was making would rule out the big public spending cuts which would be necessary if income tax were to be substantially reduced. In 1984 and 1985, the government's failures to deliver cuts in these two areas led many supporters to assume that it had lost its resolve.[20] By 1985 and 1986, in a change of tone, government spokesmen were boasting of how much they were spending on many programmes.

Against the criticisms, there is the view that her break with custom has been radical and that she has made an important difference. According to Geoffrey Smith, writing in the *Wall Street Journal* on 13 February 1985, 'Mrs Thatcher has removed some of the most familiar benchmarks of British politics. Indeed, it may be that she would be seen to have had a more lasting effect upon the pattern of British politics than of British government.'' For Peter Jenkins, she has presided over a 'Thatcher Revolution' and paved the way for a 'post-Socialist era'.[21] The 1987 government has been even more radical. Privatization has continued, household rates have been abolished and far-reaching reforms of the civil service, legal profession, and education system are under way. An admiring biographer writes that in the near future it will be impossible for any politician to be successful 'who does not accept the framework of political discussion as designed by Margaret Thatcher'.[22]

For the Marxists Hall and Jacques she has proved 'a novel and exceptional political force' and boldly challenged the mould of social democratic politics with her own beliefs in the market.[23] Some left-wing members of the Labour party view her record as a vindication of a determined, ideologically based attack on the consensus, one which they hope the forces of the left will emulate.

The background of economic decline has stimulated a radicalization of British politics and the emergence of a Thatcherite agenda is in some respects an outcome of that failure. In the space of two decades (1950–70), Britain fell from being the most prosperous major state in Europe to an economic also-ran. While other countries had their own versions of economic 'miracles', 'the only economic "miracle" was that Britain failed to take part in the progress of the rest of the world'.[24] Governments regularly came into office pledged to reverse the cycle of relative decline and each one failed. Between 1950 and 1970 the economy grew at half the rate of its Western European neighbours. In the 1970s Britain did worse on average on economic growth and inflation than other Western European states. The economic performance in this respect has to be set against the promises of British politicians and the performance of other countries.

Study of Britain's decline has almost reached the level of an industry; it has generated many conferences, editorials, books, articles, and even novels. The suggested causes are legion and some are of course contradictory—the education system, party system, civil service, non-entrepreneurial business, the dominance of the City, class divisions, bloody-minded trade unions, possession of Empire, loss of Empire, war, lack of suitable investment, too much welfare spending; or suggestions that decline was an inevitable adjustment given Britain's early start as an industrial power or that the British do not 'really' want to be competitive.

In fact Britain has long been a slow-growth economy and the growth record of the economy between 1945 and 1975 was as good as that in any period in British history. Much of the literature on decline errs in giving the impression of a past golden age.[25] Relative economic decline and the assumption of 'failure' appeared to be feeding back into the political system. The complex ways in which voters responded to economic performance are discussed in Chapter 5.

Changes in the Political Agenda

I suggest that the change in the agenda of British politics has been the outcome of three broad factors; the climate of opinion, events, and political actors. It is worth discussing each of these factors in more detail.

The climate of opinion is rather more than what is often called 'political ideas' or 'public opinion'. Public opinion in the sense of an aggregate of individual views is probably too vague a term and, in practice, is too shifting on many issues. Opinion surveys have frequently found marked fluctuations in expressed opinions on issues of the day, largely because the reactions given to pollsters may reflect weakly held views on the questions asked. The term 'climate of opinion' was used in the 1930s by an American historian, Carl Becker, to refer to an outlook or set of assumptions about policy that is largely taken for granted among the politically influential.

The climate is important in suggesting criteria for evaluating existing policies—for example, the need for a comprehensive welfare programme after 1945 or the later view that it was failing—and in suggesting changes which will realize the new values. There was a greater tolerance or permissiveness regarding private behaviour which was reflected in legislation which made for easier divorce and the decriminalization of abortion and of private sexual acts between consenting adult homosexuals in the late 1960s.

It is not claimed that there is a one-to-one relationship between a political climate and political practice. At times, indeed, ideas may be 'myths', honoured as much in the breach as in the observance and, at times, reduced to little more than slogans. Some ideas are rationalizations or defences of what politicians have done: for instance, an incomes policy becomes 'fair' and 'socialist' when practised by a Labour government. For Herbert Morrison, socialism was 'what the Labour Government does'. Also, the climate of opinion is not a monolith; it contains a mix of values, although there may be a bias in a particular direction. Political parties and interest groups have their own traditions, interests, and, sometimes, well-prepared policies. But they also respond to events (such as the energy crisis in the 1970s), reports of Commissions and working parties, research findings, and the initiatives of policy brokers. In a famous passage in his *General Theory*, Keynes claimed: 'I am sure that the power of vested interests is vastly exaggerated compared with the gradual encroachments of ideas.' In fact, however, ideas have to combine with circumstances. The phrase, 'an idea whose time has come' pays tribute not only to the strength of the idea but to the change in circumstances which has enhanced its appeal and relevance.

A second factor is *events* which impinge on the political scene. Events and trends are analytically distinct from the climate of opinion, although there is often a mutual pattern of influence between them. Events may alter public and élite perceptions of policies. At a particular point, with the German occupation of Prague in March 1939, Neville Chamberlain's policy of appeasement towards Hitler lost credibility as an effective or honourable way of containing German aggression. The rash of strikes and disruption of services in the so-called 'Winter of Discontent' in Britain in 1979 destroyed the Labour Government's claims to have a special relationship with the trade unions and the claims of the union members (as opposed to the union leaders) to owe a particular loyalty to the Labour party. The issue of what to do about the power of the trade unions returned, albeit temporarily, to the top of the political agenda. In the United States the achievement of the USSR in launching the Sputnik in 1957 was a catalyst in persuading policy-makers that the United States must invest more resources in education and in space research. Similarly, the quadrupling of the Arab oil prices in 1973–4 gave a hearing to advocates of energy conservation and promoters of alternative sources of energy. The slow-down in economic growth from the mid-1970s has ensured a more attentive hearing for those who want to contain the costs of welfare and other areas of public policy or find alternative non-state sources of financing the benefits.

The third factor covers *political actors and their goals*. Here I am referring to changes in the composition of élites, together with the manœuvres, preferences, calculations, and motives that produce particular patterns of behaviour. How explain the long-term decline of the British Liberal party without taking into account Gladstone's sudden conversion to Home Rule for Ireland in 1885 and Joseph Chamberlain's decision to break away from the party, or the decisions of Asquith and Lloyd George in 1916 which split the party? How explain the creation of the Fifth Republic in France without taking account of the role of General De Gaulle? The replacement of Mr Heath as Conservative leader by Mrs Thatcher in 1975, the election of Mr Foot as Labour leader in 1980, the decision of Mr Benn to spearhead the campaign for reform of the Labour party's constitution between 1979 and 1981 are all important factors in any convincing explanation of the shifts within the two major political parties in the 1980s. To call attention to the preferences and actions of actors in key positions is not to devalue the importance of circumstances or constraints. It is simply to remind ourselves that, where there is a margin for choice, preferences of actors have to be considered. The relationship between the personality and ideas of actors and circumstances is one of push and pull.[26] Some situations may be so loosely structured and the skills of the actor so relevant and his personality so deeply involved, that the actor plays a major role. History has many examples of individuals who are able to give history 'a push': Lenin in 1917, Churchill in 1940, De Gaulle in 1958, and, I suggest, Margaret Thatcher in the 1980s.

The three factors have so far been discussed separately. But it is important to establish some connections between them, what an American political scientist, John Kingdon, in his study of policy agendas, calls 'coupling'.[27] Kingdon explains the setting of government agendas as the outcome of three broad forces. First, there are conditions which are recognized as *problems*, such as, for example, a disaster, a crisis, or a dramatic event. Second, there is *politics* or public mood, which may be conservative or liberal, in favour of more spending on public services or in favour of tax cuts, and which arises out of the interplay of interest groups, elections, and political parties. Third, there are *participants*, for example, legislators, officials, and policy-brokers. There is no one pattern of development. Events may lead to a more receptive audience for previously unfashionable ideas and, in turn, generate 'doable' policies which are taken up by influential politicians. A spokesman for previously unfashionable ideas may achieve a new prominence. Enoch Powell in 1968 achieved some notoriety with his speeches calling for an end of immigration and repatriation for immigrants from New Commonwealth states. The speeches received immense media coverage and heightened the salience of the issue. The interaction of ideas and events creates new power situations, and what Kingdon calls 'a window of opportunity' for political actors.

Some ideas exist for a number of years, floating around in what Kingdon calls 'a policy soup'. Some survive the period of gestation but many do not. Those that do survive must meet tests of technical feasibility (they can be implemented) and value acceptability (they can gain the acceptability of the public and policy specialists). Although a number of policy-makers favoured a shift to comprehensive or non-selective secondary education in Britain as early as the 1950s, it was the conjunction of a number of circumstances in the 1960s that gave the comprehensive plan a prominent place on the agenda. These circumstances included: beliefs that such education would be a useful form of social and (eventually) economic investment and would produce a more egalitarian society; a more expansionist mood towards public expenditure in the mid-1960s; the growth in the school-age population and the disappointment of many parents at the failure of their children to achieve grammar school places; and research which suggested that there was a large wastage of talent under the existing scheme. The return of a Labour government with a large parliamentary majority in 1966 was another facilitating factor.

This book argues that the interaction of the three factors —events, ideas, and actors—explains the shift in the political agenda in Britain. The background is the continuing relative economic decline in Britain, and such events as the defeat of the Conservative party in the two elections of 1974, the collapse of incomes policies in 1974 and again in 1979, the spiralling inflation in 1975, the unpopular strikes culminating in the 1979 Winter of Discontent, and the electoral problems of both the Conservative and Labour parties. This background was coupled with a search for new policies within both parties and a disenchantment with a number of established values and leaders. In 1979 Conservative leaders, pointing to the industrial disruption and roaring inflation of the previous years, increasingly turned away from incomes policies and state ownership and the pursuit of full employment through reflation. They canvassed policy alternatives of monetarism, the free market, curbing the trade unions, and reducing the public sector. But there was also a rethinking in the Labour party, as the left gained control of policy. It advocated policies of greater state ownership and control in manufacturing and finance and British withdrawal from the European Community. More radical ideas and policies were being promoted by both the Labour left and the Conservative right. Both sides used the language of crisis. The new policies in both parties received a more sympathetic hearing, partly because the older ones had been tried and were thought to have failed, and partly because prominent politicians became associated with them. This is particularly the case with the Conservative party, in which Mrs Thatcher and Keith Joseph were sympathetic to many ideas of the New Right and became dominant figures within the party.

Plan of the Book

The contents of the subsequent chapters may be described briefly. In Chapter 2 I analyse the main features of the post-war political consensus. Consensus, as admitted already, is not an ideal term. But there was a broad agreement between the front benches of both political parties about the mixed economy and the welfare state. The chapter examines the impact of the war and its collectivist consequences in planning and welfare. In particular I single out six particular planks of the post-war policy settlement: welfare, full employment, the conciliation of trade unions, active government, optimism about the possibilities for social engineering, and the mixed economy.

Chapter 3 examines the role of groups and policy-brokers who worked to challenge and undermine the consensus. The groups on the New Right did not just attack such features as high public spending and taxes and the defects of nationalized industries, trade unions, various central and local government regulations, and the welfare state. They also produced their own policy suggestions—for education vouchers, privatization, deregulation, freeports, and so on. In general there was a greater interest in libertarian, market, monetarist, and non-governmental solutions and policies and a greater willingness to work out policy blueprints. If the Webbs, Keynes, and Beveridge were the formative figures for the social democratic policies of 1945–75, then Sir Keith Joseph, Milton Friedman, the IEA, and the CPS are the main figures of the right-wing reaction to them. These ideas and groups were helpful in supporting the neo-liberal wing of the Conservative party and it, in turn, became much more influential after 1975. Chapter 4 argues that these groups have helped to produce a New Right agenda in the 1980s.

In Chapter 5 I trace the breakdown of the consensus and organize the analysis around the six areas outlined in Chapter 2. In many, but not all, respects the social democratic settlement broke down in the 1980s.

The story of the breakdown is inseparable from the activities of certain political actors and the major political parties. Chapters 6 and 7 examine the factors that led the Labour and Conservative parties to shift, respectively, to the left and to the right. The electoral collapse of the Conservatives in 1974 and the election of Mrs Thatcher as leader in 1975 were two factors which produced a shift in the Conservative party. Similarly the collapse of the social contract in 1978–9, the constitutional reforms in 1980–1, and the exit of a number of right-wing Labour MPs to the newly-formed Social Democrats all weakened the traditionally dominant group in the Labour party.

Chapter 8 examines the record of the Thatcher government across a number of fields since 1979. It does not deal with every area of policy, although Mrs Thatcher has views—usually strongly held—on virtually all issues. It reviews the government's policies in four important areas in which it has claimed to strike out in radically different directions—welfare, industrial relations, privatization, and the economy.

Chapter 9 presents a study of Mrs Thatcher herself. She has probably been the most controversial and radical leader in the past sixty years, unusually forceful and directive, a mobilizer, and impatient with many aspects of the status quo. She has given her name to a political style and set of ideas—'Thatcherite' and 'Thatcherism'—and illustrates the importance of personality and leadership in politics.

Chapter 10 addresses the questions whether the government has made a difference to the political agenda, to public opinion, or to the powers of different groups. It also tries to relate the above changes to what has been happening in a number of other western states. There are echoes of Thatcherism in the United States, France, West Germany, and Scandinavia.

Notes

1. Quoted in R. Rose and G. Peters, *Can Government Go Bankrupt?* (London, Macmillan, 1978), p. 16.
2. R. E. Tyrrell, jun., *The Future That Doesn't Work* (New York, Doubleday & Co., 1977).
3. See P. Jackson, *Implementing Government Initiatives: The Thatcher Administration 1979–1983* (London, RIPA, 1985); P. Riddell, *The Thatcher Government* (Oxford, Blackwell, 1987); M. Holmes, *The Thatcher Government 1979–1983* (London, Wheatsheaf, 1985).

4. S. Hall and M. Jacques (eds.), *The Politics of Thatcherism* (London, Lawrence and Wishart, 1983). Also see R. Levitas (ed.), *The Ideology of the New Right* (London, Polity Press, 1986) and B. Jessop *et al. Thatcherism* (Oxford, Polity Press, 1988).
5. I. Gilmour, *Britain Can Work* (Oxford, Martin Robertson, 1983).
6. H. Young, *One of Us* (London, Macmillan, 1989).
7. Cited in D. Kavanagh, 'On Writing Contemporary History', *Electoral Studies* (1982).
8. V. Bogdanor, 'Lions and Ostriches', *Encounter* (July 1983).
9. More recently, see P. Jenkins, *The Thatcher Revolution: the Post-Socialist Era* (London, Cape, 1988).
10. S. Pinto-Duschinsky, 'Manifestos, Speeches, and the Doctrine of the Mandate', in H. Penniman (ed.), *Britain at the Polls 1979* (Washington, DC, American Enterprise Institute, 1981).
11. *Politics of the Corporate Economy* (Oxford, Martin Robertson, 1979).
12. See S. Beer, *Modern British Politics* (London, Faber, 1965) and J. Hayward, 'Political Inertia', *European Journal of Political Research* (1974).
13. B. Chapman, *British Government Observed* (London, Allen and Unwin, 1962), p. 61.
14. R. Rose, *The Problem of Party Government* (London, Macmillan, 1974), p. 124.
15. D. Kavanagh, 'From Gentlemen to Players', in W. Gwynn and R. Rose (eds.), *Britain: Progress and Decline* (London, Macmillan, 1980), p. 91.
16. *The New Conservatism* (London, Conservative Political Centre, 1980), p. 7.
17. A. Gamble, *The Free Economy and the Strong State* (London, Macmillan, 1988).
18. *Britain Without Oil* (London, Penguin, 1983), p. 19.
19. *The Thatcher Government*. See also R. Rose, *Do Parties Make A Difference?* (London, Macmillan, 2nd edn. 1984).
20. See below, Chap. 10.
21. *The Thatcher Revolution*.
22. P. Cosgrave, *Thatcher: The First Term* (London, Bodley Head, 1985), p. 224.
23. *The Politics of Thatcherism*, p. 9.
24. S. Pollard, *The Wasting of the British Economy* (London, Croom Helm, 1981), p. 6.
25. For example, M. Weiner, *English Culture and the Decline of the Industrial Spirit* (Cambridge University Press, 1981).

26. F. Greenstein, *Personality and Politics* (Chicago, Markham, 1969).
27. J. Kingdon, *Agendas, Alternatives and Public Policies* (Boston, Little and Brown, 1984).

2

THE MAKING OF CONSENSUS POLITICS

BECAUSE one may conscientiously trace the origins of many events back almost to infinity it is difficult to suggest a certain date at which modern British politics begins. It may still be possible for a student to follow a syllabus on modern British history which begins in 1485. Other plausible starting dates might be 1832, 1867, 1906, 1918, 1940, 1945, and, for some zealots, 1979.

Perhaps the best approach is to look for the emergence of such politically 'modern' features as programmatic political parties (from the 1880s onwards), large-scale bureaucracy (1914–18), regular consultation between the central government and the major interests (1916–18), a mass electorate (achieved in 1918), and acceptance of a role by the government in providing welfare and managing the economy (1930s). If we accept these as criteria of political modernity then they emerge in the years approximately between the late nineteenth century and the First World War. Before 1914 women and many men did not have the vote; there was no nationwide Labour party; the political agenda was dominated by such issues as Irish Home-Rule, powers of the hereditary House of Lords, and votes for women. The civil service was extremely modest in size and public spending amounted to only 12 per cent of GNP in 1913 (it escalated to 51 per cent of GNP in 1918 under the pressures of war). I suggest, therefore, that we date modern British politics from around 1918 or so.

The nature of British politics changed in the years that followed the 1914–18 war. Before 1914 there had been something like deadlock in the political system as the 'fundamental' issues of nationalism, state borders, and the nature of the regime polarized the two major political parties. Removal of the pre-war constitutional conflicts over the powers of the House of Lords, votes for women and Ireland eased the agenda. Although the two main parties had divided over

Home Rule for Ireland since 1886 the inbuilt Conservative majority in the House of Lords precluded progress on the issue until 1911 when the Lords' veto was reduced to two years. However, the formal withdrawal from the United Kingdom in 1920 of the twenty-six Irish (largely Catholic and rural) counties greatly simplified British society in terms of nationality, class, and religion. This event also coincided with the last great realignment of the party system. From the early 1920s onwards it was essentially a Labour versus Conservative party system and the larger electorate was more clearly divided along lines of social class. It is most unlikely that the Labour party or trade union leaders ever wanted to overturn the political system. In fact, leaders like Ramsay MacDonald, Arthur Henderson, J. R. Clynes, and Philip Snowden were more concerned to gain acceptance by the British political establishment than to overthrow it. This is not to say that governments were not seriously alarmed by outbursts of working-class unrest, particularly in 1919, 1920, and 1926; the point is that these were not led or even encouraged by the political leaders of Labour.

The received image of governmental and policy continuity in Britain does have some basis in fact. Between the 1931 election, which saw the defeat of the second minority Labour government, and 1945 there was no change of government which produced a parliamentary majority (and some guarantee of a lengthy period of office) for one party. And in general elections from 1945 to 1970 there were changes in the political complexion of government only in 1951 and 1964.

Britain's experience differed from many of her Western European neighbours in the inter-war years, particularly in maintaining such relative social and political stability. Many of these latter states faced acute problems of legitimacy in the inter-war years. Italy was still a 'new' state and had only recently been recognized by the most powerful social institution in the country—the Roman Catholic Church. The corruption and instability of successive governments were met with indifference and cynicism by the bulk of the population. In France the Third Republic tottered along; governments were short-lived and the many parties were themselves bedevilled by factionalism. In Germany many of the élite, particularly in the army and bureaucracy, looked back to pre-1914 as a golden age—and were embittered by the military defeat. Social Democrats, in power after some fifty years of opposition, were regarded by many as 'outsiders'. Instability, violence, extremism of left and right, and lack of overarching loyalty to the regime were the characteristics of politics in much of continental Europe in the inter-war years. In comparison and despite the economic problems and distress of the late 1920s and 1930s, Britain was a model of political stability and social peace.

Explanations for Political Stability

There are various explanations for Britain's political stability. One view singles out for credit the role of Stanley Baldwin and modern Conservatives in the 1920s. The private and rarely articulated concern of the small world of English 'high politics' was 'What to do about the red menace?'.[1] The Conservative leaders, it is often said, inducted the Labour Party into the ways of the British constitution, and Conservative moderation begot a similar Labour response. In December 1923, after the general election which he had called, Stanley Baldwin found that his party had a plurality (258), though not a majority, of seats. If Labour (191 seats) could count on the support of the Liberals (158 seats) it could form a minority government. Despite panic-stricken appeals to save the country from Bolshevism, Baldwin was prepared to give Labour a chance, reasoning that a minority Labour government would not imperil the constitution. Except for a few lapses he was a classic reconciler of political differences in Parliament and in the country and his party dominated government in the inter-war years.[2]

This view joins with the analyses of those critics, usually on the left, of the 'labourism' or gradualism of the Labour party. They claim that the leaders of the trade unions and the Labour party, partly by organizing and constitutionalizing working-class demands, partly by relying on Parliament, partly by negotiating and bargaining with employers, and partly by acting 'responsibly', have tamed genuine radical pressures from below. The great achievement of capitalist democracy in Britain, in this view, has been to transmute and deradicalize working-class pressures for change.[3] In Italy, France, and Germany, by contrast, there were bitter ideological divisions as parties of the left and right denied each other legitimacy and produced exaggerated responses.

Another view, which draws on the experience of Lloyd George's wartime premiership, claims that the war-government, faced by the need to mobilize organized labour, at last accepted the trade unions as an estate of the realm. In 1915 many trade unions had already agreed to give up the right to strike, accept government arbitration of disputes, and suspend restrictive practices for the duration of the war. British governments recognized the powers of unions and business, gradually sucked them into Whitehall, and appointed their spokesmen to various government advisory and consultative committees. The British style of decision-making came to resemble a form of corporatism in which the government bargained with the leaders of the major incorporated interests. Keith Middlemas claims that, over time, the role and influence of Parliament and the political parties have declined and those of senior civil servants and the producer interests have grown. The economic interests became part of what Middlemas terms 'the governing institutions', and 'these interest groups crossed the political threshold and became part of the extended state'.[4] It was a case not so much of the aristocratic or bureaucratic embrace, but of the consultative hug weakening pressures from outside for radical change.

It was inevitably a conservative approach as change could usually only be introduced with the consent of each of the partners. Some students talk of this form of consultation and bargaining as neo-corporatism, to distinguish it from the state-directed form in fascist Italy and elsewhere in Europe. Middlemas argues that the system was largely successful in avoiding crisis until the 1960s, by which time the country's continued relative economic decline increased pressures on government to tackle the powers of the interest groups.

A third and perhaps more widely accepted view dates the pattern of élite agreement back to the experience of coalition government between 1940 and 1945.[5] As ever, the stresses of mobilizing society for total war altered popular and élite perceptions towards many established policies. With national survival at stake, shortcomings in the economy and society had to be tackled urgently—usually by state action. Max Beloff has contrasted the impact of the 1914–18 war with the delays of 'normal' peacetime politics. The latter involves a period of time for preparing or 'softening up' public opinion for a new line of policy, bargaining with interests, proceeding through the legislative cycle in Parliament, and then allowing time for adjustment by administrators and the public to the new policy. By contrast, during the war, 'it was as though there was a major acceleration in the entire historical process'.[6]

An immediate effect of the 1939–45 war—as of 1914–18—was to enhance the power of organized labour. This was recognized and then furthered by the appointment of Ernest Bevin, General Secretary of the Transport and General Workers' Union, as Minister of Labour with a seat in the Cabinet in 1940. He established Joint Production Committees, on which employers and trade unionists sat, and a Joint Consultative Committee. The trade unions were drawn into the work of government and came to share many of the coalition government's assumptions about the national interest. Many of these industry-wide committees continued to operate after the war.

Another change in response to war was the notion of an implied social contract between government and people for the creation of a better post-war Britain. This had originally developed in the First World War, and was expressed in the cry 'Homes fit for heroes' and the creation of a Ministry of Reconstruction. According to Titmuss, the author of the official History of the War volume on social policy, public opinion became more egalitarian and collectivist under the impact of war and evacuation.[7] Greater awareness of social evils and poverty placed the government under pressure to create something like the welfare state. José Harris, however, claims that the Titmuss thesis may overstate the extent to which the circumstances of war actually provided a stable basis for post-war policy.[8] Many ministers were fearful of the financial consequences of a hasty acceptance of the Beveridge proposals, and there is little evidence that the war made ministers and civil servants appreciate the role of welfare as a device for promoting social solidarity or economic efficiency. Only in February 1943 did a backbench revolt in Parliament make the government accept the Beveridge package. Either way, the notion of an implicit contract for 'better times' in return for the sacrifice endured in war was widely believed.

Particularly important in reflecting the convergence between the political parties was the work of the Cabinet Committee on Reconstruction, established in 1943. This was chaired by Lord Woolton and contained Sir John Anderson, R. A. Butler, and Oliver Lyttelton for the Conservatives and Clement Attlee, Herbert Morrison, Arthur Greenwood, and Ernest Bevin for Labour. Churchill was mainly concerned with the war effort and the Labour party probably had more influence on domestic policies. This Committee reached broad agreement in many areas, covering national health service, regional policy, full employment, social insurance, and housing. It did not, however, agree on the future nationalization of basic industries.

The coalition government produced the 1944 *Employment White Paper*, passed the Education Act in 1944, the Family Allowances Act in 1945, and made progress on other aspects of the Beveridge proposals and social insurance. The government also issued reports on the future of the Bank of England, coal, gas, and electricity industries. There is some justification for Addison stating that 'the new consensus fell, like a branch of ripe plums, into the lap of Mr Attlee'.[9] The Second World War and the boost it gave to ideas of state-management, control of the economy, and welfare was a good example of how an outstanding event can alter the expectations of policy-makers, particularly their perceptions of what is politically and administratively possible. Keith Middlemas notes that 'slowly but inevitably the state came to be seen as something vaster and more beneficent than the political parties'.[10]

One has to guard against regarding the experience of the war as an all-purpose explanation for the major discontinuities in policies before and after the war. Obviously there was a massive reduction in the scale of unemployment and an increase in public spending. For most of the inter-war years state spending rarely exceeded 25 per cent of GNP; since 1945 it has never fallen below 36.5 per cent. Peacock and Wiseman have explained the post-war increases in taxation and state spending as a 'displacement effect', in which public opinion adjusts, under the impact of war, to higher levels of both.[11] Pelling, however, is sceptical of the degree of influence attributable to the war.[12] Social welfare spending increased steadily before 1939 and a number of studies have established a correlation between such spending and a country's level of economic development. Other trends, such as government intervention in the economy, consultation with major economic interests, and extensions of welfare, were, as noted, apparent before 1939 and were making headway in states which were not deeply involved in the war (for example Sweden and New Zealand). On balance, however, I am more impressed by the awareness of contemporaries and the view of many historians that the war gave a 'push' to a number of existing trends.

If the circumstances of total war facilitated the acceptance of these policies, then it is also true that the ideas on which they were based had made some progress before 1939.[13] Indeed, Peter Clarke regards the more state-interventionist Liberals of pre-1914 as the precursors of the social democracy of the 1940s—'the work of the Attlee government turned the hopeful proposals of men like Hobson, Hobhouse, and Wallas into concrete achievements'.[14] In the 1930s the failure of capitalism and the National Government to provide for the reversal of mass unemployment encouraged the development of a middle opinion across the political parties, one that believed that capitalism could be reformed. It was also a decade in which ideas about the mixed economy and welfare state were germinating. The group Political and Economic Planning established the Next Five Years' Group and Harold Macmillan published his *Middle Way* in 1938. These were signs of the emergence of a progressive centre, distinct from the political left or right, which, as Addison notes, 'came into its own in 1940'.[15] This is a fourth view which emphasizes the pre-war origins of many post-war policies.

The received wisdom about the 1930s is that it was the age of political pygmies who excluded Winston Churchill and Lloyd George from office, of mass unemployment and economic depression in South Wales, Scotland, and Northern England, and of an appeasement policy which left Britain ill prepared for war. More recently an opposite view has been presented by revisionists who point to the economic recovery and industrial adaptation achieved in these years.[16] The government provided some protection and helped the reorganization of industries, adopted a regional policy, and maintained a relatively generous welfare system. The political institutions and culture proved strong enough to prevent dynamic but politically 'unreliable' leaders like Mosley, Lloyd George, or Churchill from breaking in, and there was little support for political extremism. The contrast with the breakdown of regimes in Spain, Portugal, Italy, Germany, and Austria was clear. Critics of the general performance of the British political system in the 1920s and 1930s might pay more attention to what was happening elsewhere.

I am not suggesting that the above four views are mutually exclusive or that the historians cited would necessarily be willing to identify with one particular view to the exclusion of all others. Although there are differences of emphasis—because the views concentrate on different aspects—they do converge in explaining the shift to modern British politics. The first is about the world of high or élite politics and how policy and tactical differences were managed at parliamentary and Cabinet levels. The second concerns the interactions between leaders of central government and the economic interests in handling and preventing economic conflicts from exploding in the political arena. The third view is also about high politics, this time concerning the formation of much post-war policy in the wartime coalition government with a view to anticipating popular demands, and heeding the 'lessons' of war. The fourth and final view traces the origins of ideas which bore fruit in policies during and after the war.

Perhaps one of the most significant features of the war experience was in its failure to promote any radical change in political institutions. One might point to Churchill's tenure of office as a further instalment in the concentration of power in the hands of the Prime Minister (but not on issues of post-war reconstruction) and the decline of the House of Commons. But in terms of the major institutions—Parliament, Cabinet, monarchy, civil service, and the constitutional conventions and statutes which governed their working—the overwhelming impression is of continuity between, say, 1935 and 1950. Victory vindicated the political system. In West Germany, Italy, France, and Japan, not only were new constitutions established, but the élites, political parties, and institutions were greatly changed. For better or worse the outcome of the war actually had a conservatizing effect on the British political system.

Post-War Consensus

By the early 1950s it was already possible to refer to a post-war consensus about many, though not all, policies. In foreign affairs there developed a general agreement about such issues as Britain's role as a nuclear power, membership of NATO, and decolonization. For the first half of the century Britain had administered a far-flung empire. In the thirty years after the granting of independence to India and Pakistan in 1947 more than thirty colonies eventually achieved independence. Until 1961 there was front-bench agreement that Britain should stand aside from the European Community. There were divisions over Suez in 1956 and strong opposition in Labour to nuclear weapons and membership of the Community. But the continuity in policy between governments was impressive. In spite of Conservative rhetoric about Britain's role as a world military power and Labour's about Britain's moral leadership, the country's political leaders gracefully acceded to its loss of status.

The package of policies on the domestic front is familiar: full employment budgets; the greater acceptance—even conciliation—of the trade unions, whose bargaining position was strengthened by an increased membership and full employment; public ownership of the basic or monopoly services and industries; state provision of social welfare, requiring in turn high public expenditure and taxation; and economic management of a sort, via a large public sector and a reduced role for the market. This is the vocabulary, as it were, of modern capitalism and social democracy.

The different elements in this domestic policy package were

in many ways interconnected. The emergence of universal welfare provision depended in part on the achievement of full employment so that the welfare system would not be overloaded. Both Beveridge and Keynes envisaged a level of unemployment well in excess of that which was to be the norm over the following thirty years. The ideas of Keynes legitimized a large public sector, active government, and welfare expenditure. All were instruments of active government and enabled government to regulate demand. As Hugh Heclo remarks, 'spending for social purposes could serve economic purposes of increased production, investment, and stable fiscal policy. Social policy was not only good economics, but the economic and social spheres of public policy were integrally related with each other.'[17] The trade unions favoured welfare, full employment, and public ownership. Keynesian techniques of economic management, economic planning, and state provision of welfare all gave an opportunity for influence on a larger scale to possessors of expertise—economists, the professions, and social engineers.

One has to guard against viewing the post-war period as politically 'normal'. Between 1918 and 1945 there was a great deal of electoral instability, with the emergence of a much larger electorate and new issues. The Lloyd George wartime coalition in 1916 was both a cause and consequence of the Liberal split. It broke up in 1921 and was replaced by a Conservative government which was shorn of many of its leading figures. This was followed by a short-lived minority Labour government in 1924, a Conservative government (1924–9), another minority Labour government (1929–31), and then a coalition or largely Conservative government in the 1930s. But after 1945 two major parties alternated in office and the lines of the new, post-war settlement were clear. Between 1918 and 1945 the country had moved from a predominantly Conservative–Liberal party system to a Conservative–Labour one. The instability in the inter-war years in the parties' electoral fortunes and the volatility in voting behaviour were symptoms of the realignment process. In the 1945, 1950, and 1951 general elections, class alignment reached its peak and many voters saw major differences between the political parties.[18] For the ten years or so after

1945 there was great stability in party support, as reflected in opinion polls, by-elections and the slight shifts in public opinion during general elections. This stability in voting behaviour has never been restored in the past twenty years.

TABLE 2.1 *Voters' Perceptions of Difference or Similarities between the Parties, 1951–1979*

	1951	1955	1959	1964	1966	1970	Feb 1974	Oct 1974	1977	1979
					(in per cent)					
Are important differences	71	74	66	59	55	54	57	54	34	54
Much of a muchness	20	20	29	32	37	41	38	41	60	41
Don't know	9	6	5	9	8	5	5	5	6	5

Source: Gallup

Surveys by Gallup during each election found a steady increase, until 1964, in the proportion of voters regarding the political parties as 'much of a muchness' and agreeing that it did not make a great deal of difference which party won the general election (Table 2.1). David Robertson's analysis of the major themes in the Conservative and Labour election manifestos from 1924 to 1966 found that the two parties had converged towards the centre by the mid-1960s. By 1964 the distance between the two parties on major economic issues was only one-fifth that of 1931, when the parties were at extremes (Figure 2.1).[19] This development appeared to confirm Anthony Downs's analysis, advanced in *An Economic Theory of Democracy*.[20] Downs assumed that, on a left–right spectrum, there is a position at which parties can maximize their votes. This is not necessarily at the mid-point of the spectrum, for voters may bunch at any point on it. Downs argued that the two evenly matched parties in a largely consensual electorate would converge in their policies to win the decisive votes of the floating voters.

Apart from the survey evidence of consensus and the formal theory of convergence, some commentators also argued that there *should* be a large measure of consensus and policy continuity between the parties because modern government was

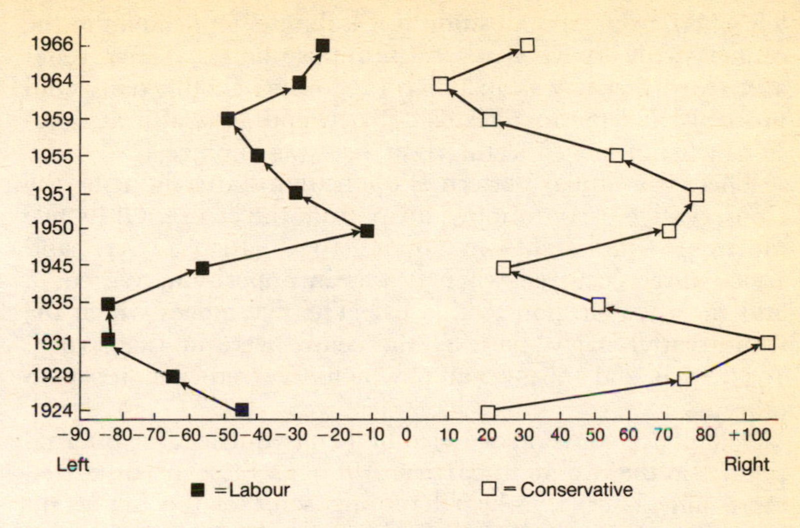

Figure 2.1. The Greater Agreement between Parties on Economic Issues 1924–1966
Source: Adapted from David Robertson, *A Theory of Party Competition* (London: Wiley, 1976), p. 98.

interventionist and engaged in promoting far-reaching social and economic objectives. According to Karl Mannheim, 'The reduction of the political element is essential to any form of planning . . . The task of straightening out its (trade cycle's) booms and slumps, and repairing the damage it has done, is only partly a political problem; it is largely a matter of science and technique'.[21] Parties and elections, in other words, should not make too much of a difference.

It is worth noting, however, that whether a party was in government or in opposition was an important influencing factor on whether it moved in an adversary (left, or more state control and nationalization, for Labour, right for Conservative) or a consensus direction.[22] For the ten general elections from 1924 to 1966, we see, in Robertson's analysis, that Labour moved to the left (compared to the previous election) five times—in 1929, 1931, 1951, 1955, and 1959—and it was in opposition on all of these occasions except 1951. We would almost certainly add the manifestos of February 1974 and 1983 to the list of occasions when Labour also moved to the left when it was in opposition. Of Labour's four moves to the centre, two were when it was in opposition (1945 and 1964) and two when it was in office (1950 and 1966). We would probably add the manifestos of 1970 and 1979 also as occasions when it moved to the right and was in office.

There is a similar pattern of opposition status inducing the Conservative party to move away from the centre. Of its five moves to the right—in 1929, 1931, 1950, 1951, and 1966—three occurred when it was in opposition. We could add the 1979 election to this list. The five moves which the Conservative party made to the centre were in 1935, 1945, 1955, 1959, and 1964—each of which occurred when it was in office.

The policy convergence also had consequences for internal politics in the two main parties. Although these are discussed more fully later, it is worth noting some of the important outcomes. In the Conservative party defenders of active government and welfare became dominant and the neo-liberal wing was effectively routed by the late 1940s. The Conservative manifestos in 1950 and 1951 called for a halt to further nationalization rather than a reversal (except for the cases of iron and steel and road haulage) and a repudiation of 'political theory' as a guide to economic policy. In other words, the party could live with most of the nationalization measures of the 1945 Labour government. Similarly, full employment was accepted as a major objective of economic policy. In 1956 Prime Minister Eden wrote to Harold Macmillan, then Chancellor of the Exchequer, about his worry over rising prices and wages. In considering ways to cope with the pressure of trade unions for higher wages, he noted the possible role of unemployment in weakening their bargaining power. However, he went on to dismiss this as being 'politically not tolerable'.[23] Macmillan, a rather isolated Conservative prophet of the middle way in the 1930s—neither socialist nor capitalist—was able to look back as Prime Minister in 1958 on how much of his programme had been accomplished.

In much of the Labour party leadership there was a more pragmatic attitude towards public ownership, explicitly under Hugh Gaitskell, covertly under Harold Wilson. Socialism, or collectivism, was overtaking capitalism, as J. A. Schumpeter in 1942 had forecast.[24] But, notwithstanding Karl Marx, it was coming peacefully. In many large firms, ownership was divorced from managerial control; managers in large firms had mixed views about the free market, distrusting the unregulated competition and unpredictability associated with it, and distinctions between public and private enterprise increasingly became blurred. Socialism for many in the Labour leadership appeared increasingly to be based more on demand management and the promotion of economic efficiency than on state ownership and directive economic planning.

There was more of a pragmatic acceptance of the policies of the mixed economy and welfare state than a principled agreement about them and in both parties there were dissenters. Conservatives could remain critical about the extent and methods of economic planning and controls as practised by Labour, and indeed the Conservative party did not turn to economic planning until 1961. Similarly within the Labour party there was always a large minority that pushed for bolder policies of public ownership.

One should not overstress the uniqueness of the British experience in this respect. In many other western states so-called 'catch-all', voter-orientated, political parties sought support from most sections of the electorate. Parties relied increasingly on public relations techniques and played down distinctive sectional and ideological appeals at election time.[25] For the general election in 1959 the Conservatives hired an advertising agency to market their campaign. In 1960 Harold Macmillan had written to a party official: 'Who are the middle classes? What do they want? How can we give it to them?' (Private information.) After the 1959 election defeat the Labour leader, Hugh Gaitskell, also realized that the process of social change among the working class meant that Labour would have to update its policies and image. By 1964 Labour had turned to private opinion polling to help it present its policies.

In the scope of a chapter one cannot deal fully with all the major planks of the post-war agenda. Among the most significant, however, were the following; full employment, mixed economy, active government, welfare, conciliation of the trade unions, and expertise.

Full Employment

Conscious pursuit of full employment as a goal of economic policy was first expressed in the famous 1944 White Paper on Employment (Cmnd. 6527). In this, the government accepted 'as one of the primary aims and responsibilities the maintenance of a high and stable level of employment after the war'.[26] Other western governments gave similar undertakings. In the United States, for example, the 1946 Employment Act required the federal government to have policies for full employment. In Britain, as noted, the policy also had a practical aspect; it was important for maintaining the insurance basis of the welfare state. The goal was accepted by all parties. The 1950 Conservative manifesto stated: 'We regard the achievement of full employment as the first aim of a Conservative Government.' Between 1948 and 1970 the annual average level of unemployment never exceeded 3 per cent, compared with the norm of 10 per cent during the inter-war years. Perhaps no other change presented such a contrast between pre- and post-war Britain.

The central figure in this story of full employment is John Maynard Keynes. Keynes was a polymath, gifted in philosophy, economics, finance, and the arts. Although he never held elective office he always seemed to be hovering near the centre of political affairs between 1919 and 1945. To acquire influence he used his pen, his membership of government and Liberal party committees, and his access to ministers and Treasury officials. Until 1940 he was like a one-man opposition to the financial and economic establishment of the day. At first Keynes's main weapon was his pen. His attack on the Versailles Peace Treaty and his polemic on Churchill's decision to return to the gold standard at $4.80 in 1925, as well as his co-authorship of the Liberal Yellow Book Proposals in 1928, were all notable publishing *coups*. He took an instrumental view of the political parties which, according to D. E. Moggridge, he used 'as vehicles for his ideas' and detached himself from them when he judged that they proved unhelpful.[27] He was, says Moggridge, a 'demi-semi-official'.

Keynesian techniques of economic management reconciled government control over the economy with political freedom.

The great choice of the inter-war years appeared to be planning versus freedom, both in the political and economic realms. Keynes was the genius who reconciled them, by showing politicians how to curb the risks and uncertainties of the market. As a tool of economic management Keynesianism could lend itself to the interpretations of both Labour and Conservatives. The economic depression showed that the classical economics of the market were not working as theorists surmised they would. Classical theories of supply and demand held that these two forces would balance to provide full employment, a theory aptly summed up in Say's Law that supply creates its own demand. If there was unemployment, then it was because wages were too high. In his *General Theory* Keynes argued that there was no necessary link between savings, investment, and consumption and gave government a role in stabilizing economic activity; government expenditure and its effects on investment should be set with the goal of promoting an appropriate level of aggregate demand. If demand was deficient and the national income and level of employment fell below otherwise possible levels the Treasury should budget for a deficit and encourage state and private spending via tax cuts, and/or reduced interest rates. If demand was too high then it should budget for a surplus by reversing these measures.

It is a matter of debate whether Keynesianism was responsible for the full employment which lasted until the mid-1970s. Some would argue that it was the product of increased investment and a long economic 'boom', from which many countries benefited. Research by Matthews showed that British governments had in fact usually been deflationary in fiscal policy and had run an average surplus on the current account of some 3 per cent of national income each year.[28] And the well-known 'stop-go' cycles of economic management were influenced less by the goal of full employment than by concern over the state of sterling and the balance of payments. Because of the Bretton Woods system of fixed exchange rates against the dollar, the priority of the government was to maintain the parity of the currency. Foreign pressures on the value of sterling were an incentive to pursue 'prudent' financial policies and practise monetary discipline.[29]

Mixed Economy

Between 1945 and 1951 the coal, rail transport, civil aviation, iron and steel, road passenger and freight transport, electricity, and gas industries were taken into public ownership. This was the outcome of the process of rationalization or 'cartelization' which had affected many industries in the 1930s and of the wartime regime of planning, control, and consultation. The idea of public control of such undertakings had already been established before the war in such public corporations as the Port of London Authority, British Broadcasting Corporation, Central Electricity Generating Board, and London Passenger Transport Board. Most acts of nationalization involved the state taking over industries, reorganizing them, and providing investment, for some of them were badly run down and required heavy investment. The Labour party justified its measures on pragmatic rather than ideological grounds. In *Public Ownership: The Next Step* (1948), the party laid down five conditions which would make an industry suitable for nationalization: inefficiency, for example poor management or low investment, bad industrial relations, monopoly position, investment of capital on a large scale, and being a basic supplier of raw materials. The industries were established as statutory monopolies, run as public corporations, and instructed to 'break even' taking one year with another, rather than make profits. There was little that was distinctively socialist about the programme.

Conservatives had not originally welcomed the nationalization measures after 1945. But, returned to office in 1951, the party accepted all the measures, except for iron and steel which it denationalized in 1953 (later renationalized by Labour in 1967). Most of the electorate seemed to have accepted the other measures. But although the Conservative party was committed to private enterprise it had never been anti-statist. One can quote many statements to this effect:

> Modern Conservatism inherits the traditions of Toryism which are favourable to the activity and authority of the state.[30]

> We are not the party of unbridled, brutal capitalism . . . we are not the political children of the *laissez-faire* school.[31]

By 1974 the above sectors were still in public ownership and being added to, notwithstanding Conservative dislike of them. The Heath government rescued Rolls–Royce and the Upper Clyde Shipyard firm by taking them into public ownership in 1972, having earlier announced that they would not rescue lame ducks. The party's *Campaign Guide* 1974, made clear its acceptance of the status quo and dismissal of denationalization: 'Conservatives much prefer private enterprise to run a high proportion of the economy but it is simply not feasible to sell off all the nationalised industries regardless of price or consequences. In many instances, the industries are natural monopolies and must remain subject to public control.' In 1975 the Labour government stepped in to buy a share in the ailing Chrysler car company, bought a 95 per cent share in the Leyland car company, and then transferred its holdings to a newly formed National Enterprise Board. The same government also took into public ownership the aerospace and shipbuilding industries, both of which were already receiving substantial state aid.

Active Government

Keynes obviously provided an important justification for active government in economic management. The greater role of government as employer, taxer, and distributor of benefits also seemed to be a useful tool for mobilizing popular consent. In a study of several western states, Douglas Hibbs demonstrated that countries which had highly socialized patterns of consumption and distribution and a large social wage (or state-provided benefits) had fewer strikes and industrial disputes than those countries which ranked low on these features.[32] Processing issues through the political arena rather than the free market seemed to produce greater social peace, at least in the 1950s and 1960s. In Britain as elsewhere, the main indicators of active government were the mixed economy and the welfare state.

Governments also gradually came to assume a commitment to promote economic growth. In 1954 the Chancellor of the Exchequer, R. A. Butler, pledged the Conservative government to doubling the standard of living in the following twenty-five years. There was some scepticism at the time, but the pledge was fulfilled. In the early 1960s governments moved to economic planning, establishing targets for economic growth, consulting with the two sides of industry and even, in 1965, establishing a National Plan. In his classic work *Modern Capitalism*, published in 1965,[33] Andrew Shonfield took for granted the continuation of steady economic growth, full employment, and the provision of comprehensive social welfare in western states. From 1959 onwards political parties competed on their ability to make the economy grow faster and then distribute the fiscal dividend of economic growth. Economic growth was a means of providing both social welfare and protecting the take-home pay of workers. From 1959 public spending also grew, on the assumption of future economic growth. But even when GNP failed to grow by the anticipated amount public spending was rarely cut back in proportion. The result was that by 1975 public spending, as then measured, amounted to nearly 60 per cent of GNP. In the twenty years after 1953 state spending was increasingly devoted to welfare goods and services, such as pensions,

TABLE 2.2 *Changes in Programmes of Public Expenditure 1953–1973* (as % of total public spending)

	1953	1963	1973	Change
Social Welfare (including social security, education, NHS, personal social services)	32.0	42.0	44.0	+12.0
Defining Concerns (including military, law and order, debt, finance, external affairs)	45.0	36.0	27.0	−18.0
Economic programmes (Housing and environment, transport, employment, agriculture, food)	21.0	20.0	27.0	+6.0
Miscellaneous	2.0	2.0	2.0	0.0
Total public spending as % of GDP	36.9	34.9	40.6	+3.7

Source: R. Rose, *Politics in England* (London, Faber, 1985), p. 395.

education, and health. In the same period spending on 'core' activities, like defence and law and order, actually fell as a proportion of total state expenditure (Table 2.2).

The most notable attempt to break with the syndrome of interventionist government was Mr Heath's administration. It speedily abolished the Prices and Incomes Board and the Land Commission, wound up Labour's Industrial Reorganization Corporation, repealed the Industrial Expansion Act, and phased out the Regional Employment Premium. Tax cuts were announced and the rate of increase in public spending was also cut back. By 1973, however, not only had the spending cuts been reversed, but the government had poured public funds into ailing firms and taken control of incomes, and its 1972 Industry Act granted such extensive interventionist powers that Mr Benn, as Labour's Industry minister in 1974–5, was able to extend state control over manufacturing firms, without passing any new enabling legislation.

Welfare

During the war there was a greater acceptance among policy-makers of the idea that citizenship encompassed a range of social as well as political rights. Sir William Beveridge's review of social security, *Social Insurance and Allied Services* (Cmnd. 6404), in 1942 had a great impact on MPs and the public. Beveridge wanted to construct a scheme to combat 'Want, Disease, Ignorance, Squalor, and Idleness'. His report proposed that existing separate schemes for pensions, unemployment, and sickness benefits be consolidated into a universal national insurance scheme. Instead of a means test, flat-rate benefits would be paid as of right in return for flat-rate contributions. In addition, there would be a 'safety-net' of means-tested national assistance benefits for those not covered by social insurance. Beveridge envisaged that social security should be part of a comprehensive plan for welfare and include a national health service and full employment. It is worth noting, however, that the report retained an important strand of individualism which has subsequently been seized on by members of Mrs Thatcher's Cabinet. Beveridge wrote: 'The Welfare State should not stifle incentive, opportunity, responsibility, in establishing a national minimum. It should leave room and encouragement for voluntary action by each individual to provide more than that minimum for himself and his family.'[34] In 1946 family allowances were made available from general taxation. In the same year a National Insurance Act provided flat-rate benefits for those insured, covering unemployment, sickness, retirement, and widowhood. In 1948 the National Assistance Act was brought in to cover those who did not have a complete contributions record. In 1947 the National Health Service was established to provide free medical services for all at the point of entry.

These main features of the welfare state were largely accepted by the Conservative governments in the 1950s. The welfare state acquired an ideological life of its own, incorporating ideas of fairness, a common society, and collectivism.[35] We can talk about the existence of a large measure of inter-party agreement on the welfare state which prevailed at least until 1979. Yet as early as 1965 some Conservatives, perhaps liberated by the experience of opposition for the first time in over thirteen years, were rethinking the principle of universality. They looked more favourably on the idea of selectivity, or the concentration of aid on those in real need. The idea of selectivity received a bad press because it evoked memories of the hated, pre-war means tests. At the same time, although social expenditure grew from 18 to 29 per cent of GNP between 1961 and 1975[36] many people across the political spectrum were critical of the shortcomings of the welfare state.

A large part of the increase in such expenditure was facilitated by the sharp fall in defence spending, from a quarter to one-tenth of government spending between 1953 and 1973. Yet welfare expenditure was also increasing rapidly in other western states during these years, partly as a response to economic growth. Interestingly, the highest spenders were Denmark, Norway, and Sweden—with predominantly socialist governments, but societies which also enjoyed the highest living standards in Western Europe. Peter Flora argues that the original stimulus for establishing many welfare measures was largely defensive, as Liberal or Conservative political leaders adopted these measures in part to ward off pressure for more fundamental economic and political reform from working-class parties and movements.[37] Hugh Heclo notes that hardly anywhere did the working-class movement initiate or enthusiastically support the early social insurance principles.[38] The policies usually originated from policy experts and middlemen (such as the Webbs, Beveridge, or Titmuss) who believed they knew what was good for the workers. In recent years, however, the welfare constituency has expanded beyond the poor to include much of the middle class.

Welfare was also linked to broader economic policy. A clear illustration of the connection was seen in the Social Contract drawn up between the TUC and the Labour party in 1973. In this the TUC promised to exercise restraint in wage negotiations in return for a variety of welfare measures, including food subsidies, price controls, rent controls, and an increase in pensions. The 1975 Employment Protection Act covered the right not to be unfairly dismissed, redundancy pay, maternity rights, and minimum periods of notice. Yet many observers and trade union leaders were aware that welfare benefits, in so far as they were provided out of higher taxation, also stimulated pressure for wage increases.

Conciliation of Trade Unions

The trade unions soon turned away from their brief flirtation with direct action in 1921 and 1926. More significant for the future were the abortive talks between Mond (an industrialist and former Minister of Health) and Turner (Chairman of the TUC) in 1928 to explore ways of increasing co-operation between trade unions and employers. After the collapse of the 1926 general strike most unions were in any case too weak to take industrial action. They continued to regard Labour as 'their' party and wanted a Labour government because this was a way of ensuring favourable legislation on industrial relations and social policies; both Labour and the unions wanted the repeal of the Trade Disputes Act of 1927, which replaced the contracting-out principle of making payments to the union's political fund by contracting in, and barred civil servants from undertaking sympathy strikes and civil service unions from affiliating to a political party.

Within the Labour party the major trade unions acted at

the party's annual Conference as a hammer of the left in the 1930s and again in the 1950s. Yet there was tension between the unions and the Labour party. Some Labour leaders, in their desire to show their 'fitness' to govern, calculated that they had to appeal beyond the trade unions and indeed at times stand up to them. At the same time the Trades Union Congress had to work with governments of both parties, if it was to bargain for its members. By tradition the party was run according to a separation of spheres between the two wings of the movement; the trade union leaders left largely political matters to the parliamentarians and they expected to be left alone in wage bargaining and industrial relations. Although there was a period of severe wage restraint between 1948 and 1950, political interference in these two areas was regarded as 'off limits' within the Labour movement. Both Keynes and Beveridge had been worried about the potential inflationary effects of free collective bargaining in a situation of near full employment. Keynes claimed that this was an 'essentially political problem'. The quid pro quo of a full employment policy was that the unions would exercise restraint in wage negotiations. The 1944 White Paper stated: 'If . . . we are to operate with success a policy for maintaining a high and stable level of employment, it will be essential that employers and workers should exercise moderation in wage matters.'[39] In so far as incomes restraint was ruled out for the first half of the post-war period we can say that the consensus, represented by the ideas of Beveridge and Keynes, was a selective one.

The Labour government operated a voluntary policy from 1948 with some success until 1950. Succeeding Conservative governments relied on exhortation, warnings, and occasionally restraining public-sector groups as an example to the workforce. The combination of steadily improving living standards during the 1950s and a deliberate policy of conciliation of the unions at the Ministry of Labour helped to avoid any major confrontation between government and the unions. Incomes policies as an answer to wage-induced inflation arrived on the agenda in 1961 with Selwyn Lloyd's 'pay pause'. Macmillan wanted to couple a search for ways of stimulating faster growth and economic planning, represented by the establishment of NEDC, with lower wages settlements.

For the latter purpose a newly established National Incomes Commission would set appropriate norms. But the TUC would have nothing to do with the latter body and it had a short life.

The consensus on the trade unions was an uneasy one. Successive Labour and Conservative governments felt pressures to intervene and impose limitations on the unions' bargaining rights as an answer to the inflationary consequences of free collective bargaining and to unofficial strikes (Table 2.3). The first Wilson government had been elected in 1964 on a platform of economic modernization. In the 'new' dynamic economy which Labour promised to create there would be no place, according to Harold Wilson, for restrictive practices: the Prime Minister urged the AEUW Annual Conference in 1966 that 'your rule book (be) consigned to the industrial museum'. But restraint on incomes soon emerged at the forefront of policy. In 1966 the Labour government established a National Board for Prices and Incomes to which proposed wage increases were to be referred and which had the power to impose a three-month standstill. This was soon followed by a phase of statutory controls which lasted until 1968.

Even more significant was the appointment in 1965 of a Royal Commission under Lord Donovan to inquire into the unions' and employers' organizations. To the government's disappointment its report in 1968 was largely non-interventionist in its recommendations. Notwithstanding the report the government in 1969 produced its own proposals to limit the unions' bargaining rights, largely because of growing concern over the increasing number of unofficial strikes. It wanted to wind down the incomes policy but take some steps which would impress foreign holders of sterling. The proposals included a twenty-eight-day cooling-off period before strikes took place, pre-strike ballots, and penal sanctions against unfair industrial practices. As a sweetener there was also a series of measures favourable to workers and unions. The proposals were of course abandoned in the face of widespread opposition among unions and the party in Parliament, including the Cabinet. The new Conservative government produced its own legislation to regulate collective bargaining in

TABLE 2.3 *Summary of UK Incomes Policies 1948–1976*

Period	Name	Government	Voluntary/ Compulsory	TUC Co-operation	Institutions
Feb 48–Oct 50	Cripps-TUC	Labour	Voluntary	Yes	None
Jul 61–Mar 62	Selwyn Lloyd's pay pause	Conservative	Voluntary but imposed in public sector	No	None
Apr 62–Oct 64	Guiding light	Conservative	Voluntary	No, refused to co-operate with NIC	National Incomes Commission (NIC)
Dec 64–Jul 66	Statement of Intent	Labour	Voluntary	Yes	National Board for Prices & Incomes (NBPI)
Jul 66–Dec 66	Freeze	Labour	Statutory	Acquiescence	NBPI retained
Jan 67–Jun 67	Severe restraint	Labour	Statutory	Acquiescence	NBPI retained
Jun 67–Apr 68	Relaxation	Labour	Statutory	Acquiescence	NBPI retained
Apr 68–Jun 70	Jenkins: renewed restraint	Labour	Statutory	Acquiescence	NBPI retained
Nov 72–Jan 73	Stage I Freeze	Conservative	Statutory	Hostile compliance	
Feb 73–Oct 73	Stage II	Conservative	Statutory	Hostile compliance	Pay Board Price Commission
Nov 73–Feb 74	Stage III	Conservative	Statutory	Hostile compliance	Pay Board Price Commission
Mar 74–Jul 74	Social Contract	Labour	Voluntary		
Aug 75–Jul 76	£6	Labour	Compulsory (not statutory)	Yes	None
Aug 76–Jul 77	4½%	Labour	Compulsory (not statutory)	Yes	None

A Summary of UK Incomes Policies 1948–76 (continued)

eriod	Wage norm	Actual * wage increases†	Actual price increases†	Associated conditions and concessions	How ended
eb 48– ct 50	None	2.4%	3.1%	(i) prices on controlled goods frozen (ii) dividends frozen (iii) voluntary price and profit restraint by FBI etc.	TUC Congress voted to abandon wage restraint
ıl 61– Iar 62	Zero for new agreements	4.3%	4.6%	None	Breached by Electricity Council in November 1961
pr 62– ct 64	2–2½% p.a. adjusted to 3½% p.a. in 1963	4.3%	2.7%	(i) 'Neddy' indicative planning apparatus (ii) 4% growth rate	Faded away
ec 64– ıl 66	3–3½% p.a.	7.4%	4.2%	(i) National plan (ii) 4% growth target	'Blown off course' by seamen's strike May/June 1966
ıl 66– ec 76	Zero, roll back of previous agreements	0.1%	3.5%		
ın 67– ın 67	'Continued restraint'	8.6%	4.9%		
ın 67– pr 68	3½% plus productivity agreements raised to 3½–4½% at end 1969	7.1%	5.4%	(i) Abandonment of 'In place of Strife'	'Dirty Jobs' pay explosion 1969/70
ov 72– ın 73	Zero	1.1%	7.3%	(i) Effective non-implementation of Industrial Relations Act (ii) 5% growth target?	

ncrease in the index of basic hourly wage rages †At annual rate

A Summary of UK Incomes Policies 1948–76 (continued)

Period	Wage norm	Actual* wage increases†	Actual price increases†	Associated conditions and concessions	How ended
Feb 73–Oct 73	£1 per week plus 4%	14.1%	11.0%	(iii)Subsidies to State industries	
Nov 73–Feb 74	7% plus partial indexation	12.8%	18.9%		Miners dispute and February election defeat
Mar 74–Jul 75	Wages to move in line with cost of living index	32.0%	24.4%	Repeat of Industrial Relations Act, food subsidy, gift tax, etc.	Sterling crisis provokes compulsory policy
Aug 75–Jul 76	£6 per week flat rate	17.5%	12.9%	(i) Renewed commitment to egalitarian plans	
Aug 76–Jul 77	£2.50–£4 per week			(i) Reductions in tax (ii) Nationalization of Aircraft and Shipbuilding industries	

Source: From S. Brittan and P. Lilley. *The Delusion of Incomes Policy* pp., 154–55
*Increases in the index of basic hourly wage rates
†At annual rate

the Industrial Relations Act and also moved to a statutory incomes policy in 1972. Interestingly, Labour had tried to legislate in the sphere of industrial relations as a replacement for incomes policy; the Conservative government reversed the timing of the two policies. The Industrial Relations Act was consciously designed as an alternative to incomes policy. The 1972 U-turn to incomes policy was closely connected to the destruction of the Act.

The traditional approach to industrial relations had been clearly expressed in the Donovan Report. The Report was largely the work of industrial relations specialists who were either Oxford academics or had some connection with that university. They firmly believed that the introduction of legally binding collective agreements was not the answer to unofficial strikes or other forms of unofficial action. This group was in favour of 'voluntary' action, for, according to the Report, 'The British system of industrial relations is based on voluntarily agreed rules which, as a matter of principle, are not enforced by law.' The Wilson and Heath governments rejected this philosophy and thought that the law not only had a role to play but was indispensable in reforming industrial relations. The paradox was that the failure of the Wilson attempt to introduce legislation and the ineffectiveness and political and industrial costs of the Heath legislation only appeared to confirm the validity of the voluntarist case.

Successive governments came to view trade unions as a major barrier to their policies of arresting the country's relative economic decline and attacking inflation. Many politicians considered the trade union question to be central in British politics from 1964 onwards. Incomes policies were variously statutory or non-statutory, compulsory or voluntary, and were often linked with conditions and concessions about social and economic policy by the government of the day. They were also accompanied by an array of commissions and boards to provide guidance on particular claims and deals. Although the policies produced a temporary slow-down in inflation, in every case where there was a target set for the average level of pay the out-turn, as measured by increased basic hourly wages, exceeded the target and by the second or third year the policy collapsed. As in 1950 it was usually caused by a failure of trade union leaders to carry their delegates, activists, and members. The lack of central authority within the TUC or within many unions made it difficult to operate a long-term policy. Incomes policies ended in grief and damaged relations between government and union leaders and between the latter and their members. The resort to an incomes policy was invariably justified by an alleged economic crisis—high inflation or the weakness of sterling. In no case did a party in government promise in advance of election that it would have such a policy.

Periods of Labour government since 1948 have been almost inseparable from incomes policy. At times the policy has been dressed up as an essential part of socialism, with reference to redistribution and economic planning. Indeed, governments were prepared to consult and bargain, notably over the Social Contract in 1974–5, and in 1977 the Chancellor of the Exchequer, Denis Healey, linked tax cuts in his budget with an anticipation of moderation in wage settlements. Free collective bargaining was criticized as a form of group coercion and a format which rewarded the strong and penalized the weak. Some might see irony in the explicit abandonment of market principles in the field of incomes alone, given the absence of shared agreement over differentials and other issues. Under the Social Contract the trade union movement gained much favourable legislation from the Labour government and was in regular and close touch with it. This period 1974–5 seemed to be the height of the new group politics.

In the 1950s and 1960s only a few ministers or officials accepted a monetary explanation of inflation. And, if they did, they thought the cost—in the form of higher unemployment following from a monetary squeeze—would be politically and electorally unacceptable. In 1958 the Chancellor of the Exchequer and his junior ministers had resigned when Mr Macmillan's Cabinet refused to support an extra £50 million in spending reductions. The departing ministers also wanted a tighter control on money supply.[40] The tendency to invoke incomes policy as an answer to inflation stemmed from the intellectual climate. Brittan and Lilley state:

> The British opinion-forming classes—civil servants, politicians, commentators and academics—had largely stopped thinking in terms of the market mechanism. They felt much more at home with politically determined 'strategies'. Consequently the market was readily assumed to have 'failed' even when it was working, and when inflation did worsen this was always attributed to the inherent weaknesses of collective bargaining rather than to prior monetary excess.[41]

Expertise

There was general optimism that many of the widely shared goals in social policy could be achieved and a belief that the relevant knowledge for social engineering was also available. After all the government had managed to achieve full employment so why could it not produce similar improvements in the fields of education, housing, poverty, regional policy, and so on? Professor David Donnison has recently commented on the assumptions about welfare of Crosland and the Titmuss school at the London School of Economics in the 1950s.[42] He notes that there was a general belief among the policy-making élite that the provision of the social wage and progressive taxation were ways of softening inequalities of income. Economic growth would provide the resources for a relatively painless shift towards a more equal society. Governments (usually Labour) would take the initiative aided by the public service professions, the Conservatives would largely acquiesce in policy innovations (compare Sir Keith Joseph and the ratchet),[43] and what Donnison calls 'middle England' would approve. In the field of education, for example, academic studies pointed to a great waste of talent which could profit from further education.[44] The concern was met by an expansion of education, particularly in higher education, and the shift to comprehensive secondary education. By the 1970s the commitment to greater equality along a range of issues, such as race and gender, had extended to the establishment of the Race Relations Commission, the Equal Opportunities Commission, and the Royal Commission on the Distribution of Income and Wealth (1974–9).

Another strand of the belief in expertise was seen in the vogue for managerialism, incomes policies, and economic planning in the 1960s. The Public Expenditure Survey Committee system was introduced in 1961 and planned totals of public spending for five years ahead. In 1962 a National Economic Development Committee was established and based its work in 1963 on an economic growth target of 4 per cent per annum. In the 1964 general election the parties competed on which of them could make the economy grow fastest. Michael Stewart notes that the activist political leadership of the time had little sympathy for relying on the price mechanism. For Harold Wilson, 'If resources were to flow to where they were most needed—to exports and industrial investment—they had to be pushed there by "direct intervention", by physical controls or official and unofficial arm-twisting. What was needed, if British industry was to be modernised and to compete in the world, were improvements in management, more industrial training, the development of new techniques and their embodiment in new investment.'[45] His government was committed to economic planning and established a Department of Economic Affairs which could argue against the Treasury for economic expansion. In 1965 the DEA drew up a plan which proposed policies for government intervention in industry, training, and manpower with the objective of achieving a growth rate of 25 per cent between 1964 and 1970. In July 1966, faced with a major sterling crisis, the government took severe deflationary measures and aborted the plan.

The shift to economic planning gave a greater role to experts and the bureaucrats and also encouraged calls for continuity between the political parties.[46] Trevor Smith has noted:

> Indeed, as was so often stated in the great debate (about planning), the role of consensus was to depoliticize those areas of policy (mainly economic and industrial) where, it was believed, a free-ranging application of appropriate skills would be the most effective means of achieving prosperity.[47]

The mood was also reflected in the institutional reformism of both Mr Heath and Mr Wilson between 1964 and 1976. Local government was reformed, a new system of select committees was established in Parliament, the Central Policy Review Staff and a Policy Unit for the Prime Minister were established, and attempts were made to reform the trade unions, introduce devolution for Scotland and Wales and alter the House of Lords. A poignant example of this belief was seen in the Fulton Report on the Civil Service (1968) which advocated the recruitment of administrators with 'relevant' skills, particularly those in the social sciences. The criteria of relevance were never spelled out.

No doubt the reformers hoped that institutional change would be a catalyst for other changes. But institutional change, largely because it is easier to bring about than economic growth or substantive policy changes, may easily be a form of displacement activity. Such reforms were, as Johnson observes, often 'advocated as an alternative to the much more difficult task of modifying attitudes and behaviour in the society, which are themselves very serious obstacles to achieving the desired changes'.[48]

Conclusion

Much of this consensus existed at the élite level and did not necessarily reflect popular attitudes (indeed critics argued that one of the chief virtues of modern representative democracy was that it kept 'demos' at bay). 'Reasonableness' came from on high. Opinion leaders in the mass media, front-bench politicians, senior civil servants, and many pressure group leaders thought that the above policies represented the lowest common denominator. Yet surveys showed no great support for many of the 'core issues' in the consensus—public ownership, welfare benefits for the undeserving poor, abolition of capital punishment, some practices of trade unions, decolonization, and, at times, British membership of the European Community (by contrast Atlanticism was favoured both at élite and popular level); and, perhaps above all, the belief that all British subjects (some 700 million) should have unrestricted entry to Britain.[49] The importance of Enoch Powell's speech about immigration in 1968 (see p. 71) was that a senior 'insider' broke out of the high-mindedness of the consensus. For a brief period he was the most popular politician in the land. Powell used mass fears to attack élite attitudes—the classic populist strategy—and in so doing he showed that parts of the consensus rested on unsteady foundations.

There is no gainsaying that the first thirty post-war years of consensus politics coincided with both a steady reduction in Britain's international standing and a relative economic decline. Equally, one has to acknowledge that in every post-war year until 1973 [except 1958] the economy grew: the period saw the most rapid rise in living standards and social welfare. In this sense, the consensus coincided with economic success, even though Britain's economic performance was failing to match that of most other western states. Defenders of the consensus may also object that evidence of relative economic decline dates back over a century and clearly predates the era of consensus politics. Moreover, other countries, notably Sweden, Austria, and Norway, pursued consensual social democratic policies and had outstanding economic records.

Governments in many Western European states sought moderation in wage settlements from organized labour and other producer groups in return for government policies covering unemployment, welfare expenditure, and taxation. These 'tripartite' or 'social partners' arrangements worked well enough in the 1950s and 1960s in such states as West Germany, Sweden, Norway, the Netherlands, and Austria, all of which had low levels of inflation and unemployment. These countries usually had centralized trade union confederations which were able to bargain authoritatively for their members and a strong Social Democratic (or a strong political left) presence in government.

There was also a significant decline in strike activity in Belgium and Britain compared to the inter-war years and strikes virtually disappeared in Scandinavia, Austria, and the Netherlands.[50] In many countries labour movements appear to have held off and traded low increases in earnings for employment. There was an emphasis on seeking consensus, providing for continuity of policy, working through an array of councils, commissions, and so on, and separating as far as possible economic management from partisan considerations. France and Italy were notable exceptions to the 'social partners' approach, for the trade unions were too weak to be involved in such a process and in the United States the size of the country and diversity of interest groups ruled it out. In Britain, as noted, the 'social contract' or incomes control approach was followed episodically but usually collapsed after two or three years.

Claims about the demise of ideology between the parties and the convergence across many industrial western states not surprisingly prompted the question, 'Does Politics Matter?' According to one observer, 'This ideological agreement, which might best be described as "conservative socialism", has become *the* ideology of major parties in the developing states of Europe and America.'[51] Broadly similar policy packages were emerging in other Western European states in the same period. The growth of scientific thought and expertise, exemplified in Keynesian economics, appeared to weaken the appeal and relevance of ideologies of Left and Right. Affluence softened social and class polarization and narrowed policy differences between parties of the left and right. The role of government in public spending and state employment grew. The latter more than doubled as a share of the workforce in Italy and Sweden between 1950 and 1980.[52] The government's 'take' from citizens in direct and indirect taxes amounted to over a third of GDP, and the provision of welfare moved in similar directions in western states despite their different histories, party systems, and forms of government.[53]

It is possible to argue that the dissatisfaction so widely expressed in the 1960s and onwards signalled a break in the consensus. The attacks on traditionalism and on many prevailing assumptions in industry, the trade unions, the civil service, and Parliament were coupled with appeals for modernization and professionalism in industry, higher education, and so on. In fact, the 'What's wrong with Britain?' school was in many ways an attempt to shore up the old consensus. Rarely were the outlines of the post-war settlement, which was largely accepted among the political élite, civil service, economic interests, and mass media, called into question. The remedies were widely regarded as institutional change and a more technocratic managerial style of leadership.

There is a reasonably precise point at which a sense of relative political and economic decline developed among the political leaders. The awareness of Britain's loss of international influence was reflected in the failure of the expedition to hold the Suez Canal in 1956 and then the belated and abortive applications to join the European Community in 1961 and 1967. This was linked to the greater awareness that other Western European states were achieving much faster rates of economic growth. In 1959 the Labour opposition had tried to make some political mileage out of comparative economic growth rates, but without much success. By 1961, however, the pursuit of economic growth was a publicly stated objective of all political parties. What is remarkable is that the call for more economic planning by Labour in 1964, or for a social market approach by the Conservatives in the 1970s, involved marginal changes. Both parties were still trying to make the consensus work; the pursuit of economic growth was largely seen in terms of working for a bigger cake which could be distributed by politicians to the electorate.

The growing sense of dissatisfaction in the 1960s and 1970s (though this anticipates Chapter 5) was reflected in the appointment of many investigative Royal Commissions and committees of inquiry, and the reform of several institutions. However, the economic performance was disappointing; indeed, inflation and unemployment both rose to higher levels. In 1973 there was the first explosive increase of higher oil prices and the onset of economic recession. The governments were regularly turned out of office, ignominiously so in February 1974 and 1979, and support for alternative political parties grew. Even here, however, it is worth noting that the Liberal party, which made the greatest gain in votes, promised a return to consensus (as did the SDP in 1981). There were still votes in what Rolf Dahrendorf called a vision of A Better Yesterday.

Notes

1. M. Cowling, *The Impact of Labour* (Cambridge University Press, 1967).
2. K. Middlemas and J. Barnes, *Baldwin* (London, Weidenfeld & Nicolson, 1969).
3. For example, see R. Miliband, *Capitalist Democracy in Britain* (Oxford University Press, 1983).
4. *Politics in Industrial Society* (London, Deutsch, 1979), pp. 371–82.
5. P. Addison, *The Road to 1945* (London, Cape, 1975).
6. M. Beloff, *Wars and Welfare Britain 1914–45* (London, Edward Arnold, 1984), p. 23.
7. R. Titmuss, *Problems of Social Policy* (London, HMSO, 1950).
8. J. Harris, 'Social Planning in War-Time: Some Aspects of the Beveridge Report', in M. Winter (ed.), *War and Economic Development* (Cambridge University Press, 1975).
9. *The Road to 1945*, p. 14.
10. *Politics in Industrial Society*, p. 272.
11. A. Peacock and J. Wiseman, *The Growth of Public Expenditure in the United Kingdom* (Princeton University Press, 1961).
12. *Britain and the Second World War* (London, Collins, 1970).

13. S. Beer, *Modern British Politics* (London, Faber, 1965). P. Addison, *The Road to 1945* and H. Pelling, *The Labour Governments* (London, Macmillan, 1984).
14. P. Clarke, *Liberals and Social Democrats* (Cambridge University Press, 1979).
15. *The Road to 1945*, p. 18.
16. J. Stevenson and C. Cook, *The Slump* (London, Cape, 1978). I. Gilmour, *Britain Can Work* (Oxford, Martin Robertson, 1983).
17. 'Towards a New Welfare State', in P. Flora (ed.), *The Development of Welfare States in Europe and the United States* (New Brunswick, Transactions, 1984), p. 391.
18. D. Butler and D. Stokes, *Political Change in Britain* (London, Macmillan, 1969).
19. *A Theory of Party Competition* (London, Wiley, 1976).
20. (New York, Harper & Row, 1957.)
21. *Man and Society* (London, Kegan Paul, 1946), p. 360.
22. R. Rose, *Do Parties Make A Difference?* (London, Macmillan, 2nd edn. 1984).
23. Cited in S. Beer, *Modern British Politics*, p. 360.
24. *Capitalism, Socialism and Democracy* (London, Unwin, 1943).
25. O. Kirchheimer, 'The Transformation of Western Europe Party Systems', in J. La Palombara and M. Weiner (eds.), *Political Parties and Political Development* (Princeton University Press, 1964), and S. M. Lipset, 'The Modernisation of Contemporary European Politics', in S. M. Lipset (ed.), *Revolution and Counter-Revolution* (London, Heinemann, 1964).
26. Cmnd. 6527, p. 3.
27. D. E. Moggridge, *Keynes* (London, Fontana, 1976), p. 38.
28. R. Matthews, 'Why has Britain had Full Employment since the War?', *Economic Journal* (1968).
29. S. Brittan, *Steering the Economy* (London, Penguin, 1971), p. 93.
30. Q. Hogg, *The Case for Conservatism* (London, Penguin, 1947), p. 294.
31. Sir A. Eden, cited in S. Beer, *Modern British Politics*, p. 271.
32. D. Hibbs, 'On the Political Economy of Long-Run Trends in Strike Activity', *British Journal of Political Science* (1978).
33. (Oxford University Press.)
34. *Social Insurance and Allied Services* (London, HMSO Cmnd. 6404), pp. 6–7.
35. H. Heclo, 'Welfare: Progress and Stagnation', in W. Gwynn and R. Rose (eds.), *Britain: Progress and Decline* (London, Macmillan, 1980).
36. I. Gough, *The Political Economy of the Welfare State* (London, Macmillan, 1979), p. 76.

37. P. Flora (ed.), *The Development of Welfare States in Europe and the United States.*
38. H. Heclo, 'Towards a New Welfare State', in P. Flora (ed.), *The Development of Welfare States in Europe and the United States.*
39. Para. 49, p. 18.
40. On this, see S. Brittan, *Steering the Economy*, p. 212.
41. S. Brittan and P. Lilley, *The Delusion of Incomes Policy* (London, Temple Smith, 1977), p. 178.
42. D. Donnison, *The Politics of Poverty* (Oxford, Martin Robertson, 1982).
43. See below, p. 114 ff.
44. J. Flood, A. Halsey, and F. Martin, *Social Class and Educational Opportunity* (London, Heinemann, 1966).
45. M. Stewart, *The Jekyll and Hyde Years* (London, Dent, 1977), p. 28.
46. A. Shonfield, *Modern Capitalism*, p. 235.
47. T. Smith, *The Politics of the Corporate Economy* (London, Martin Robertson, 1979), p. 88.
48. Editorial, 'The Reorganization of Central Government', *Public Administration* (1971), pp. 1–2.
49. 'Westminster and Whitehall', in W. Gwynn and R. Rose (eds.), *Britain, Progress and Decline.*
50. D. Hibbs, 'On the Political Economy of Long-Run Trends in Strike Activity', *British Journal of Political Science* (1978).
51. S. Lipset, 'The Modernisation of Contemporary European Politics', in S. Lipset (ed.), *Revolution and Counter Revolution* (London, Heinemann, 1964), pp. 244–5.
52. R. Rose, *Understanding Big Government* (London, Sage, 1984).
53. S. Lipset, 'The Modernisation of Contemporary European Politics', and A. King, 'Ideas, Institutions and the Policies of Governments', *British Journal of Political Science* (1973).

3

THE CONSENSUS QUESTIONED

In much of political science literature there is a divorce between, on the one hand, studies of political concepts and theory and, on the other, studies of political institutions and processes. Some students indeed have disclaimed the need to relate the analysis of political thinkers and political theories to the real world, in the sense of relating the theories to the circumstances which prevailed when they were written or speculating about their possible consequences for political behaviour today. The students are more concerned with analysing a particular theory, examining the internal logic and coherence of a work, and prescribing rules of conduct. John Plamenatz, for example, in the preface to his justly praised *Man and Society*, dismisses the case for relating theories to the conditions and controversies existing at the time when they were written with the question, 'Is there to be no division of labour?'[1] He also notes that the task which most great political philosophers set themselves was more speculative than explanatory, for 'the aim was often less to explain than to justify or condemn'.[2]

Political Ideas and Practice

In contrast to Keynes's acknowledgment of the importance of ideas,[3] Marx regarded ideas and values as part of the superstructure, which in turn derived from economic interests. The dominant ideas of any epoch were, for Marx, the ideas of the ruling class. At the turn of the century a number of élitists like Pareto, Michels, and Mosca, as well as a burgeoning school of psychologists, claimed that ideas and beliefs were often covers or rationalizations for interests. No doubt it is an over-simplification to see ideas and circumstances, or even ideas and interests, as always distinct and opposed. John Stuart Mill was correct when he wrote that ideas must conspire with circumstances if they are to be successful; such circumstances include the availability of resources, political support among the élite and public opinion, and a widespread sense that other ideas had been tried and failed or, for various reasons, were not found to be politically or administratively practicable. Interests do shape ideas and ideologies—to mobilize support, legitimize demands, undermine opponents, promote solidarity, and so on. But, in turn, these ideas may either promote interests or they may be only tenuously connected to them. After all, it requires a great leap of imagination to perceive the similarity of interests as motivating the very different policies of the Heath government during 1972–4 and the Thatcher government since 1979. Values are not always derivative and apparently similar interests frequently give rise to different values.

Many case studies on the evolution of policies have emphasized the greater importance of circumstances compared to ideas. Paul Smith's study of the social reforms of Disraeli's government between 1874 and 1880, for example, downgrades the importance of the young Disraeli's ideas. Smith sees the reforms as the outcome of *ad hoc* and piecemeal decisions, 'an array of personal, departmental, and technical considerations'.[4] One of the more celebrated cases of the apparent influence of ideas has been that of the late nineteenth-century Fabians and, in particular, of Sidney and Beatrice Webb. The Webbs' strategy for promoting their policies was to permeate the established Conservative and Liberal political parties, and only when this approach failed did they turn to the infant Labour Party. They emphasized the importance of empirical research and proposing feasible policies. Beatrice was appointed to the Royal Commission on the Poor Law in 1905 and largely co-authored a significant Minority Report which advocated state provision of a comprehensive social security programme some four decades before it was to be implemented. Their tactics of permeation involved planting collectivist ideas and policies in the minds of policy-makers. As Beatrice wrote in *Our Partnership*, 'we want the things done and we don't much care which party gets the credit'.[5] Yet an authoritative historian of the Fabian Society has found that other forces were, on balance, more important than the

Fabians in the cases of the policies adopted by ILP, the Liberals, and the Labour party.[6]

David Donnison has recently reminded us of the many social scientists in Britain whose combination of empirical research and policy commitment have influenced debates on policy.[7] They include, apart from the Webbs, Booth, Rowntree, Tawney, Burt, Keynes, Bowlby, and Titmuss. To an increasing extent their latter-day equivalents are found in the social scientists in universities and research institutes.

A number of factors help the diffusion of new ideas among policy-makers. One is the small number of relevant influential policy-makers, about 100 politicians and the 300 or so officials who are regularly in contact with ministers. This rather small, homogeneous group operates in a 'closely knit world of interlocking networks and institutions', centred in London.[8] According to F. M. G. Wilson, 'They are not—cannot be—reflective moulders of ideas; they are the recipients, the collaters, the compromisers among a mass of persuasive influences and persistent people. They do not—and have not time to—think.'[9] There are also informal links between members of government and members of the Oxbridge colleges and the LSE Common Room. The small size of the policy community means that an idea can make rapid progress once it has entered the community. In Britain, however, there is no counterpart to the very large numbers of prominent academic social scientists who are brought in by a new President to occupy some of the most senior posts in the American federal government.

By repute British politicians are pragmatic, more interested in the practical than the theoretical. They are also interested in ideas which suggest an attractive line of policy or which provide a theoretical justification for particular courses of action. If Marxism has not been strong in the Labour party neither has Conservative 'theory' been significant in the Conservative party. As a rule politicians are interested in concrete ideas and they often come to the ideas second or third-hand through conversation or reading newspapers. They may turn to groups or individuals who play a role as 'brokers' or mediators between the worlds of ideas and political practice.

They are also interested in ideas which can be turned into 'do-able' policies. Politicians live in a public realm and they need ideas and phrases to persuade, explain, and justify courses of action before an audience, be it the political party, interest groups, or public opinion.

Political 'ideas' come in different forms. One form might be certain approved names and associated political values. For example, Disraeli, Winston Churchill, and Burke, or Aneurin Bevan and Keir Hardie, are frequently cited at Conservative and Labour gatherings respectively. The same is true of the invocation of such defining ideas as liberty and freedom or brotherhood and equality. The mention of these symbolic names and ideas increases the sense of togetherness and community between a speaker and his audience. Graham Wallas appreciated this when, in 1908, he wrote about the function of political parties at elections as 'something which can be loved and trusted, and which can be recognised at successive elections as being the same thing that was loved and trusted before; and a party is such a thing'.[10] A second form might be philosophical guides or objectives from which policies may be derived. For Labour extending social and economic equality, strengthening the bargaining position of trade unions, or expanding welfare provide criteria of 'good' policies. For the Conservative party such values include lower taxation, secure national defence, law and order, and free enterprise. A third form might be specific policies designed to promote the values.

Policy as Learning

We do not know a great deal about how ideas germinate in policy communities or how the climate of opinion changes and influences decision-makers. For example, A. V. Dicey seemed to take for granted that there was a harmony between legislation and the dominant tendencies of English thought in the nineteenth century. Proof of the influence of ideas was to be found in legislation and vice versa. The ideas were, he says, 'in the air', a response to the era. In the mid-nineteenth century the ideas of Bentham were influential in many policy areas, notably finance, law, penology, and administrative practice (particularly in India—a *tabula rasa* for the enthusiastic reformer). Finer has suggested three means by which the Benthamite ideas were transmitted to decision-makers. These included *irradiation*, through personal contacts at salons and clubs, *suscitation*, or stirring public opinion through the press, Royal Commissions, and parliamentary select committees, and *permeation*, through the appointment of sympathizers to commissions and committees. Finer notes the interconnection of these different methods:

> IRRADIATION made friends and influenced people. Through them, SUSCITATION proved possible. SUSCITATION led to official appointments and hence PERMEATION. And permeation led to further irradiation and suscitation: and so on *da capo*.[11]

Ideas for policy come from many sources—from pressure groups, political parties (including their traditions), annual conference resolutions and Research Departments, the senior civil service, and policy-brokers and experts from research institutes, universities, and other learned bodies. One should not, however, exaggerate the amount of innovation that takes place within the formal policy process itself. A majority of former Cabinet ministers interviewed by Bruce Headey saw themselves as legitimators of an existing policy rather than as innovators.[12] Much policy-making is rather routine, concerned with carrying out existing policies and making marginal adjustments according to changing circumstances or changing preferences, for example, establishing a new level of welfare benefits, reducing or extending the range of beneficiaries, changing the criteria of eligibility, and so on.

Some observers argue that political parties are poorly prepared in opposition and that the leadership in Parliament spends too much time reacting to what the government of the day is actually doing. The Research Departments of the parties are not well staffed or financed, and have other tasks (primarily secretarial and parliamentary services in the case of the Conservatives, formulating long-term policy for Labour) than preparing immediate policies. Opposition leaders usually have little or no contact with senior civil servants or interest groups and may often lack an awareness of the administrative practicability of the particular policies that they are developing. At the same time this very freedom may give them the opportunity to be radical and innovative.

Senior civil servants are the confidential advisers to ministers and they have a special responsibility for presenting policy options and advising on policies. Although British civil servants are their own experts on many subjects even they turn to leaders of relevant interest groups to seek advice about likely reactions to a policy and about its implementation. Political outsiders have been brought into the service, notably during wartime and on a modest scale since 1964. Yet many have soon withdrawn, claiming that the culture of Whitehall has limited their effectiveness.[13] It has often been noted that administrators are not encouraged to be radical and indeed are a brake on radical change; a probable consequence of being permanent and anonymous is that the official has little to gain or to lose from commitment to particular policies.

An important source of ideas has been policy-brokers who operate between the political and academic worlds. In Britain there is a well-established tradition of policy-orientated research. Individuals like Beveridge or Rowntree were able to point to weaknesses in existing lines of policy and pave the way for social reforms. Banting calls these individuals 'middle men', who are 'drawing upon a broader stream of research, theory and opinion, developing in the academic and professional communities'.[14] The cycle of politico-academic influence usually starts with the findings of scholarly research being reported via memoranda and conferences to fellow experts; the ideas gradually find their way to a wider audience through the quality weeklies and press; and then they percolate through Parliament and the Civil Service. Such work is important in promoting an awareness of problems, defining the nature of issues, and specifying policy alternatives. One of the best known recent examples of research changing the perceptions of policy-makers in the areas of poverty and inequality is that of the so-called 'Titmuss school' at the London School of Economics. Titmuss and colleagues like Michael Young, Peter Townsend, and Brian Abel-Smith used careful empirical research to demonstrate the failures of many existing welfare programmes in eradicating poverty and to advance their egalitarian values. Banting notes, 'Their research was explicitly political: its aim was to reshape policymakers' interpretations of their environment.'[15] Many new programmes followed in the 1960s and 1970s, including the educational priority areas, urban programmes, housing action areas, and community action programmes. Similarly, much research in education in the 1950s and 1960s cast doubt on the notion that there was a fixed pool of ability which could profit from academic secondary education.

Not all academic disciplines are equally useful to the policy-maker. To be useful a discipline requires at least (a) a scholarly consensus on its findings—even though this very agreement may foreclose policy options—and (b) a set of politically and administratively feasible instruments or methods for transforming the knowledge into 'do-able' policies.[16] Economics has been, until recently, one subject area on which policy-makers have drawn since it has been assumed to possess the above features. Theologians or sociologists *qua* practitioners of their subjects have not been in high demand.

Powellism

The survival of the Conservative party is in part due to its flexibility about matters of political doctrine. According to Michael Oakeshott political behaviour appropriately takes place within the tradition of a society. Politics is not something to be derived from premeditated ideas or an ideology; there is no Tory equivalent of *Das Kapital*. Conservatives have combined an attachment to the status quo with an acceptance of gradual reform. Change may be necessary, not least to preserve, but it should be gradual, evolutionary, and should conform to the nature of society. Conservatives do not hesitate, in the words of Lord Hailsham, 'to make use of the lessons taught by their opponents'. They see nothing immoral or even eccentric in 'catching the Whigs bathing and walking away with their clothes!'[17] The greatest obstacle to attempts to discern an 'essence' of British conservatism has been the party's 'dual inheritance', libertarianism and collectivism. During the nineteenth and twentieth centuries there have been many prominent Conservative advocates of the market economy and *laissez-faire*, as well as defenders of state intervention to manage the economy and provide welfare. But there is no doubt that in government in the twentieth century the collectivist stand has dominated the policy thinking of Conservatives.

An attempt to restate a more libertarian, free-market version of Conservatism was made by Enoch Powell in the 1960s. Powell had been a junior minister in the Treasury until he resigned in 1958, along with Peter Thorneycroft, the Chancellor, over the Macmillan Cabinet's refusal to accept the public spending cuts recommended by the Treasury. He entered the Cabinet as Minister for Health in 1961, but then refused to serve under Sir Alec Douglas Home in 1963. In opposition from 1964 he expounded a free-market, *laissez-faire* approach to the economy. His criticisms of incomes policy, economic planning, regional policy, and high levels of public spending and his advocacy of flexible exchange rates were aimed not only at the then Labour Government but also at previous Conservative Governments. Enoch Powell is perhaps the only post-war British politician of whom it is possible to write a serious study of his or her political thought. He is always concerned to derive his position on issues from fundamentals, and frequently talks about such matters as the purpose of political activity, the relationships between the individual, community, and state, and the nature of freedom.[18] The free market is, according to Powell, essential for a free society, for distributing power as widely as possible. It is 'the subtlest and most efficient system mankind has yet devised for setting effort and resources to their best economic use'.[19] Many of these views were expressed in bold, even startling language; when he spoke political observers took note. Prices and incomes policy was dismissed as 'A nonsense, a silly nonsense, a transparent nonsense. What is more and worse, it is a dangerous nonsense.' Already as a junior minister in 1958 he was explaining inflation in terms of the rapid growth of government spending and the money supply. These sentiments gained a better hearing and influenced some of the policies adopted by the party in opposition after 1966 and in the early stages of the Heath government (Chapter 7, below). But Powell, the spokesman for the market economy, was soon overtaken by Powell the British nationalist, the opponent of immigration and British membership of the European Community. His famous 'rivers of blood' speech about coloured immigration in 1968 made him a formidable rival to the party leader, Edward Heath. The popular support for Powellism was largely for his views on immigration rather than for his views on the free market.[20]

Mr Heath dismissed Powell from the Shadow Cabinet after the 1968 speech and the latter became a regular critic of the 1970 Heath government, particularly for its statutory prices and incomes policy and membership of the EEC. Powell resigned his seat when Parliament was dissolved in January 1974 and returned to the Commons the following October as Ulster Unionist MP for South Down in Ulster. Ironically, as the Tory party came round to accepting many of his economic ideas so Powell had already broken with the party.

A Silent Majority

Many of the 'New Right' figures have given the impression that they are outnumbered by spokesmen for leftist and collectivist values in the mass media. The New Right in the United States has similarly felt itself outnumbered. Joseph Schumpeter anticipated them in *Capitalism, Socialism and Democracy* when he argued that capitalism would destroy itself by creating an anti-capitalist culture.[21] He claimed that the intelligentsia, the offspring of an affluent society and expanded higher education, would be critical of entrepreneurial values, the profit motive, the free market, and materialism—and would not be employed in business. The growth of large business corporations would encourage a bureaucratic rather than entrepreneurial outlook. Businessmen would not be skilful communicators and would only be selective in their support for the values of competition and free enterprise, which they would abandon when it suited them. Schumpeter feared for capitalism also because bourgeois leaders lacked the colour and charisma of a monarchy or military or aristocratic rulers, and the values of free enterprise and profit were not calculated to inspire the electorate. In the United States an important element in the neo-conservative outlook has been the criticism of a 'new class' which is highly educated, well endowed with communicating skills, and disproportionately employed in the mass media, universities, and interest groups. Irving Kristol claims that the values of the new class are predominantly anti-business, permissive, and even anti-patriotic.[22] Kristol, however, is not uncritical of the 'free market' and has condemned its lack of a legitimating myth or a theory of social justice.

In Britain leading Conservatives, including Mrs Thatcher, have complained that Conservatives are not naturally 'political', certainly when compared with the greater commitment shown by activists on the political left. The party, they claim, had lost the initiative in schools, universities, and even churches in the 1960s and early 1970s to proponents of collectivist, egalitarian, and anti-capitalist values.[23] The role traditionally played by intellectuals in propagating values, providing authoritative interpretations of political social and economic trends, and recommending policies has increasingly been played by opinion-formers in the mass media and universities. Daniel Bell, in his influential *The Coming of Post-Industrial Society*, [24] claims that because of the importance of theoretical knowledge and innovation in such a society, the university has replaced business as the key institution. It also provides an opportunity for people to think about social and political issues and a platform for them to publicize their ideas. A survey of university teachers by Halsey and Trow found that they were indeed more left-orientated and more supportive of the Labour party than were other middle-class groups.[25] Later (see below p. 114) there was to be strong evidence of the dominant Keynesian outlook among British economists at universities. A survey of journalists by Tunstall in the late 1960s also found that many specialist journalists were to the political left of their organizations and were more likely to be pro-Labour than pro-Conservative.[26] The Conservative concern about the drift of ideas was well expressed by Ronald Butt, a sympathetic commentator in the *Sunday Times*, on 20 October 1974:

In the past decade, the whole vocabulary of political and social debate has been captured by the Left, whose ideology has fundamentally remained unanswered by the Conservatives. Where the Conservative party has answered back, it has done so by conceding half the case that it should have been rebutting and has usually sought to appease the 'trend'.

He stated that the party

> . . . needs, even at the risk of being called reactionary, to react. With the language of politics so largely monopolised by the Left and with the intellectuals whose activities have so much influence on society mostly talking that language, the Conservative party needs politicians with strong persuasive power and clear ideas who are utterly committed to the Conservatives' historic role.

But this claim by Conservatives that the opinion-formers helped to shift the middle ground in favour of the left co-existed with beliefs in the existence of a 'silent majority', which was not represented by an allegedly 'left' mass media (such as the BBC). Indeed, there was ample survey evidence that the electorate was well to the right of the major parties on issues like immigration, law and order, discipline and standards in schools, capital punishment, and trade unions; public policy in these areas was made by a parliamentary élite which did not reflect public opinion.[27] A set of populist and authoritarian policies, it was claimed, might mobilize great working class support.[28] Enoch Powell occasionally claimed to be no more than an echo of popular feelings on issues like immigration and the European Community: the people were, he asserted, deprived of a choice because of the front-bench consensus on these issues. Powell was also to claim, first on immigration and then on the issue of Ulster's relationship to Ireland, that a minority—in the civil service and in the media—was manipulating the majority. The minority so controlled the means of communication 'that the majority are reduced to a condition in which they finally mistrust their own senses and their own reason, and surrender their will to the manipulator'.[29] The voice of this silent majority could be strengthened by such policies as increasing parental influence in the management of schools, requiring more frequent elections of trade union officials and more pre-strike ballots, reducing the size of the public sector, curbing the power of powerful producer groups, and, above all, by a bolder and more distinctive right-wing leadership in the Conservative party.

That free-market ideas gained a more sympathetic hearing among Conservatives in the 1970s was in part born out of a reaction in the party to the U-turns towards interventionist economic policies by the Heath government in 1972 and to the obvious failure of those policies in the 1974 election. The ideas were nourished largely outside the official Conservative party machine and reinforced by the failures (economic and electoral) of the subsequent Labour government under Wilson and Callaghan. A historian of the Conservative party claims that Mrs Thatcher's task in 1979 was greatly assisted by 'the backing of an intellectual revolution' which Mr Heath did not have.[30] Labour Party historian Ben Pimlott states that 'ideas . . . stimulated the Tory revival . . . they came from outside, creating a groundswell of sympathetic opinion before their adoption by the Conservative party leadership'.[31] Two of the seminal thinkers in this revival of so-called New Right ideas —who were influential on Sir Keith Joseph—have been Friedrich von Hayek and Milton Friedman. In his essay on 'Intellectuals and Socialism' Hayek has stressed the importance of the production of knowledge and of the role of intellectuals, authors, academics, journalists, and teachers in shaping the climate of opinion. Many would claim that recent British experience demonstrates that ideas, and not just socialist ones, do count. (In an unpublished paper, 'Men of Affairs' (1983), Trevor Smith has drawn attention to the remarkable increase in literary activity by Conservative MPs. In the 1979 Parliament no less than a third of Tory MPs were authors of some form of publication.)

The think-tanks and lobbies were interested in changing the prevailing climate of opinion in Britain, in particular challenging the collectivist consensus and promoting more market-orientated policies. In view of the hold of the collectivist consensus this meant breaking with the status quo in many policy areas, a most un-Conservative stance. With the Conservatives in search of new policies and with Mrs Thatcher as leader of the party the groups were assured of a more sympathetic hearing than hitherto. They include such groups as the Institute of Economic Affairs (IEA), Centre for Policy Studies (CPS), Adam Smith Institute (ASI), Aims of Industry, and Institute of Directors.

These groups have looked abroad, notably to the United States, for ideas. This is particularly true of the IEA, which has promoted the monetarist ideas of Milton Friedman, the Austrian School of Economics, and the Public Choice approach of the Virginia School to economics in the United States. What these groups admire about that country is the greater encouragement it provides for free enterprise, small business, and free-market policies in welfare and education. If we may speak of an official ideology of the United States it is one that has traditionally been hostile both to a powerful central government and collectivist values.[32] Americans have been less likely to turn to the central government as a provider of education or medical care and less likely to be in trade unions. The United States is still the only industrial society without a significant socialist party. Compared to other western states the size of the public sector measured by employment and government spending as a share of GDP is much lower in the United States. If anything, the presidency of Ronald Reagan and his policies of supply-side tax cuts and deregulation have increased this admiration. But the United States has also been something of an anti-model for traditional Conservatives. After all, it was a republic which ostentatiously broke with much that the right revered—monarchy, aristocracy, hierarchy, and an established Church. It promoted the values of equality and did not invest government with any sense of majesty. The absence of a traditional ruling class, of marked social class differences, or of deference to figures of authority, and the populism and commercialism which affects so many areas of life, including politics, alienate many British and Continental Conservatives.

There is no major British writer—no equivalent of Keynes—who has emerged to inspire this New Right thinking. British figures are prominent in any Conservative pantheon, but not in this more policy-orientated, New Right collection. Groups like the IEA and politicians like Margaret Thatcher and Sir Keith Joseph have popularized (in the sense of simplified and introduced to a wider or non-specialist audience) the ideas of people like Hayek and Friedman. Accordingly we start our examination of the ideas of the British New Right with a discussion of these two seminal foreign figures.

Hayek

Hayek's distinguished work across so many disciplines—economics, philosophy, and political theory—stamps him as a genuine polymath. Recognition of his talents as an economist was reflected in the award of a Nobel Prize for Economics in 1974. There is a remarkable continuity in his work dating back to his early training in Austrian economics. He achieved some fame and notoriety with his critique of collectivism and planning in his *Road to Serfdom*.[33] Many of these ideas were developed later in *The Constitution of Liberty*[34] and three volumes of *Law, Legislation and Liberty*.[35] Nicholas Bosanquet takes the view that Hayek has become more influential in the world of affairs than in academe, partly because his warnings appear to have been largely vindicated by the eventual failures of Keynesian and full-employment policies.

Hayek is often coupled, mistakenly, with Friedman as an intellectual godfather of New Right economics. Although both are from the classical liberal tradition there are important differences between the work of the two men. Hayek claims not to be a Conservative. Conservatives, he believes, distrust the market, are content to preside over the status quo (even when it is collectivist) and are illiberal. Conservatism 'may succeed by its resistance to current tendencies in slowing down undesirable developments, but since it does not indicate another direction, it cannot prevent their continuance'.[36] Compared to Hayek, Friedman's writings are less ambitious or at least cover a more limited field. Samuel Brittan persuasively argues that Hayek is attracted to two different philosophies.[37] One is classical liberalism, with its emphasis on limited government, rule of law, and free markets; the other is conservatism, with its respect for tradition and the wisdom inherent in many existing institutions and practices.

Hayek's work can be organized around four main themes, all of which have been emphasized by the New Right. The first, most fully explicated in *Road to Serfdom*, is that central planning is evil. He objects not to modest planning but to a central plan, arguing that centralized economic planning by the government is both politically dangerous and economically inefficient. In practice it reduces individual and group liberties, upsets the balance between political institutions (by giving too much power to the executive), weakens the role of Parliament, and undermines the rule of law by granting so much discretion to authorities. Central planning invariably leads also to pressures for yet more controls as the 'logic' of intervention feeds on itself. His economic objections are that planning suppresses competition and leaves control in the hands of what are usually monopolies. Planners seek more control and predictability whereas progress comes from experiments and voluntarism; the virtue of the market is that by harnessing the knowledge it increases the opportunities for creativity and invention.

These themes have been pursued in Hayek's later writings, although the more recent targets have been government controls of incomes and policies of redistribution and welfare. It is easy to see how Hayek has been regarded as an arch-opponent of socialism and Keynesian ideas, even though *Serfdom* was written against 'the socialists in all parties' and belongs to a body of antitotalitarian writing in the late 1940s and early 1950s, which includes Karl Popper's *The Open Society and its Enemies* and J. B. Talmon's *The Origins of Totalitarian Democracy*, as well as the work of Michael Polanyi and Michael Oakeshott. But the book also attracted controversy because it challenged the belief, prevalent among parties of the moderate right and left, in a middle way in which freedom and planning could be combined. Hayek claimed that there was no compromise between 'the commercial and the military-type of society'.

A second theme in Hayek's work concerns the complexity of society. He claims that there is a spontaneous natural order which is the outcome not of a plan or of design, but of human action. Tradition, institutions, and patterns of behaviour have evolved into rules of conduct and actions of individuals and a coherence emerges from these interactions—'from each element balancing all the various factors operating on it'. Bosanquet draws attention to the almost Burkeian sympathy for the status quo in *The Constitution of Liberty*, when Hayek writes of the need for 'submission to undesigned rules and conventions whose significance and importance we largely do not understand', and that because 'man cannot know more than a tiny part of society'[38] he should guard against intervening or trying to alter these relationships. This is part of his scepticism about social engineering and relates to his attack on scientism in the study of society, particularly the study of economics. In *The Counter-Revolution of Science* Hayek emphasizes the complexity of the economic system and scorns the belief that we can predict economic behaviour on the basis of studying relationships between macro-economic variables.

A third element in Hayek's work is the importance attached to markets and prices for the allocation of resources. The spontaneous interaction of buyers and sellers is regarded as being more efficient than the activity of planners. In his more recent work he relates this theme to an attack on trade unions which are able to force an increase in money supply above the growth of production and so add to inflation. He charges that trade unions, because of their structures and practices, prevent an efficient labour market and slow down the response of the levels of wages and employment to changing market conditions.

Finally, the role of government is given an important but limited place in setting the framework for social and economic activity and upholding the rule of law. By the latter Hayek means the existence of general rules which prohibit arbitrary behaviour. It is not the job of government, according to Hayek, to promote social justice (which he regards as a vague, even arbitrary idea anyway) or, as is often the case, to compensate for and distort the working of the market. He is also critical of majoritarian democracy: too often this is a perversion of constitutionalism, a course of action being justified on the grounds that it is backed by a majority. Hayek's point is that the amount of popular support is irrelevant if the action is arbitrary. He praises ideas such as the separation of powers and the supremacy of the constitution, and argues for prior political and constitutional reforms to make the market economy and the rule of law work.

Friedman

The major spokesman for monetarism, the market economy, and the counter-attack on Keynesian economics is the

American economist Milton Friedman. Though never far from controversy, his academic reputation has been reflected in the award of a Nobel Prize (in 1977) and his election as president of the American Economic Association. Like Keynes he has publicist gifts of a high order. He has written a regular column in the American *Newsweek* magazine and his best-selling *Free to Choose* (1980) was made into a television series in Britain and the United States.

Friedman has emphasized the importance of the supply of money in causing inflation. Put simply, monetarists claim that if an increase in the money supply outstrips the growth of GDP then it will produce inflation. He made the case for a monetarist explanation of inflation in his *A Monetary History of the United States*, written with Anna Schwartz.[39] For Friedman the main cause of the Great Depression was the rapid decline in the money supply. He also believes that there is an underlying long-term equilibrium between supply and demand, given appropriate changes in money supply and exchange rates; there is, therefore, a 'natural' wage level at which people will find employment. Trade unions may distort this level and push for higher wages but only at the cost of pricing workers out of jobs. In principle, monetarist ideas are politically neutral between the left and right; after all, it was Mr Healey, the Labour Chancellor of the Exchequer, who introduced monetary targets in 1976. But because monetarists talk of a natural level of unemployment and admit that a short-term consequence of slowing down the rate of monetary growth may be an increase in unemployment, they have aroused the anger of the political left and the trade unions, and have been associated with neo-conservatism. Moreover, Friedman is also an energetic proponent of the market economy. But a monetarist economist, believing in the monetary cause of inflation, may or may not support the market economy, oppose the trade unions, high taxes, and public spending, be a 'Thatcherite', and so forth.

Friedman has also engaged himself on social as well as economic questions. He is associated with such ideas as the negative income tax, by which the poor receive cash directly rather than 'free' services, and education vouchers which people may use to 'purchase' education in schools of their own choice. In his *Capitalism and Freedom* and later in *Free to Choose*, monetarist ideas were firmly placed in a free-market context. The books present standard New Right themes—the inefficiency of government and failure of many of its programmes, benefits of lower taxes, need to denationalize and deregulate state industries and services, and abolition of protective legislation such as rent controls, minimum wages, regional and industrial subsidies, and employment legislation. These are all seen as counter-productive to the efficient working of the market economy. Friedman allows government a role in maintaining law and order and defence and providing essential services which some groups, children for example, are not able to provide for themselves. He also claims that there is a causal connection between capitalism and personal freedom, for markets disperse power and decisions whereas politics concentrates them. Freedom for the individual consists in making choices and an absence of coercion by others. Capitalism, or the voluntary interaction between buyers and sellers of goods and services, permits this economic freedom which, in turn, is essential for political freedom. But, he argues, government is also inefficient; its interventions, to protect declining industries and regions, for example, have usually failed to achieve their objectives. Not surprisingly, he has also advocated political reforms to limit the range of government. These measures include constitutional limits on taxes and public spending, a constitutional amendment to require a balanced budget, and fixed levels of monetary growth.

Institute of Economic Affairs

The Institute of Economic Affairs (IEA) was incorporated in 1957 as a 'research and educational trust' to study the role of markets and pricing in allocating resources and registering preferences. Its creation has been credited to Antony Fisher, founder of Buxted Chickens, who was persuaded by Hayek to set up a research body as the most effective way of countering the tendencies leading to 'serfdom'. The starting point of much of the IEA's work is that the climate of opinion colours the thinking of politicians. 'They are puppets on a string in the way they respond to fashions,' as the IEA's Deputy Director

John Wood expressed it.[40] For nearly thirty years the IEA has published several series of discussion papers, including Hobart Papers, and in 1980 it started its own journal, now the monthly *Economic Affairs*. Its pamphlets and monthly Hobart lunches in its Westminster offices are aimed at opinion-formers. The institute still believes in the formative role of ideas and attempts to influence the educational and academic world. In this it supports the claim of Keynes rather than Marx, even though the principal goal of its publications has been to combat much of Keynes's influence on economic thinking. The IEA has come into its own as the ideas of Keynes, Beveridge, and the Fabians—which it has challenged since the late 1950s—are increasingly in retreat, and it is now firmly established as the intellectual home of markets, liberalism, and monetarism in Britain.

The main figures in the IEA are the General Director, Lord Harris, ennobled by Mrs Thatcher in 1979, and the former Editorial Director, Arthur Seldon. The Institute makes much of not being formally attached to a political party nor financed by one. It is firmly in the libertarian tradition, and Professor W. H. Greenleaf has perceived it as the latter-day equivalent of the Anti-Socialist Union and the Liberty and Property Defence League of the early decades of this century[41] (and therefore in tune with the present Conservative party).

The IEA has played a major role in publicizing the ideas of Hayek and has published a number of his essays, including *Agenda for a Free Society* (1961), a symposium on *Constitution of Liberty*, *A Tiger by the Tail* (1972) on the trade unions, *Full Employment at any Price* (1975), and *Denationalisation of Money* (1976). The ideas of Friedman have needed less institutional publicity because of his own skills in this area, but the institute has published his 1970 Wincott lecture on *The Counter-Revolution in Monetary Theory* and *Unemployment versus Inflation* (1975).

The institute has to date drawn on various schools of economic thought to challenge Keynesianism. One is the emphasis on the market as a means of allocating resources, discovering new uses of resources, and facilitating comparison between different products and services and different providers of these goods. A second borrows from the Austrian

School of Menger and von Mises. This emphasized that economic decisions are made by individuals and that it is the marginal adjustments in prices and costs that influence decisions. The institute has been more interested in the problem of micro-economics than in macro-economics and this has probably encouraged its anti-Keynesianism. It has also brought the ideas of the Public Choice School to a wider British audience. The most famous proponents of this school have been Professors Buchanan and Tullock, formerly of the University of Virginia at Charlottesville and now of George Mason, USA. Their writings challenged the view that government or collective choice is the activity or outcome of some disinterested group. They also argue that governments, budgets, and bureaucracies grow in large part because of the ambitions of the people involved. Government decisions are usually the combined outcome of vote-seeking politicians, selfish interest groups, and expansionist bureaucrats. This theory of the political process, and of how the role of government as employer, spender, and taxer grows, fits in with some right-wing preferences for reducing the role of government and promoting the free market. The school has been less concerned with specific policy recommendations than with pointing to the negative side effects of majoritarianism. Because majoritarianism may be the enemy of public choice, the IEA has also shown some interest in institutional reforms and external constitutional limits on governments, which would give a greater scope for free markets.

The IEA, through its pamphlets, has also provided a vehicle whereby writers may make positive proposals for less government and greater competition. There is no official IEA view. (It is quick to claim that it publishes work by non-monetarists, non-marketeers, and non-Conservatives.) Much of the work starts from the belief that tax-financed services by government conceal the costs of these services, do not respond to consumer choice, and place no constraint on demands for the services. It is also critical of the government monopoly or near monopoly in the provision of many services. Among the ideas that pamphleteers have advocated have been vouchers for education, negative income tax, greater use of pricing in health and education, sale of council houses, abolition of exchange controls, an independent university (1969), opposition to universal social benefits, and the privatization, contracting-out, and deregulation of many government services, particularly in welfare. Its first pamphlet, published in 1957, questioned the state's dominance in the field of pensions. The institute has also commissioned public opinion surveys in 1963 and 1978 which, it claimed, show widespread support for its ideas in reforming the welfare state.[42]

The IEA regards itself primarily as a publishing house and the published opinions as those of individual scholars and in no way committing the institute. But there are two recurring themes across the IEA works. The first emphasizes the need for a limitation on the role of government (central and local), particularly in those areas where markets can supply services, and its concentration on providing essential public goods, such as defence or clean air. Second, there is a belief in the benefits of markets and competitive pricing. The institute also favours the approach of neo-classical political economics (which appreciates the role of individuals, pricing, and supply and demand) overturned by Keynes. Seldon and Harris write affectionately of the interaction of buyers and sellers as the democracy of the market place—'the daily referendum of the market'. Their publications are frequently critical of many aspects of the NHS and nationalized industries and complain that the state's near monopoly provision of the health and education services, combined with the avoidance of economic pricing, prevents redistribution between the poor and rich and greater help for those in the greatest need.

A. J. Culyer, in 'The IEA's Unorthodoxy', has noted that the publications deal primarily with micro-economic policy questions.[43] Culyer suggests that the lack of writers of a socialist or collectivist persuasion perhaps follows from the greater interest of most left-of-centre economists in macro-economic questions. The studies are heavily analytical, empirical, and relatively unsophisticated in terms of econometric techniques, a consequence of the aim of being widely read.

The IEA has undoubtedly played an important role in changing the climate of opinion from the mid-1970s onwards and has in turn profited from the change in climate. Enoch

Powell in the early 1960s and Sir Geoffrey Howe and Sir Keith Joseph in the 1970s were among the few senior politicians to show an interest in its work. The ideas gradually gained a wider hearing in the late 1960s and early 1970s, largely through the newspaper columns of people like Sam Brittan in the *Financial Times*, Peter Jay in *The Times*, and later William Rees Mogg, editor of *The Times*. In 1970 the institute published Friedman's Wincott lecture, *The Counter-Revolution in Monetary Theory*, which propounded the monetarist analysis. But only a few academic economists in the 1970s including Harry Johnson, Alan Walters, and Brian Griffiths at the London School of Economics, David Laidler and Michael Parkin, then at Manchester, and Patrick Minford at Liverpool challenged Keynes's ascendancy among British economists. From being so long in the wilderness, Bosanquet says the IEA began as David, has become Goliath, and now 'is in danger of becoming an orthodoxy in itself'.[44] Seldon claimed in 1984: 'Sceptics would say to us, in defence of the collectivist policies: "Give them time, they'll work in the end". By the mid-1970s they had been saying it for thirty years and could no longer ask for more time.' In challenging fashionable views across so many areas, Seldon has explicitly refused to be limited by 'considerations of current administrative practicability or political acceptability'.[45]

Social Affairs Unit

The Social Affairs Unit was founded in 1980 with intentions in social affairs similar to those of the IEA in economics. Its authors are mainly not economists but other social scientists, criminologists, philosophers, educationists, and theologians. According to one self-description the immediate programme was to build 'a systematic literature on the practical outcomes of government attempts at social engineering in the field of education, health, social welfare, discrimination, and criminal rehabilitation'. Many authors writing for the unit have been highly critical of the status quo and many of the assumptions which dominated the public sector in general and social policy and the welfare state in particular. They have questioned and often debunked the alleged expertise of various welfare professions, especially social workers and teachers, the more exaggerated analytic and predictive claims of various social sciences, again of social work, sociology of education, and criminology. A repeated theme has been scepticism about the old confident optimism in social engineering. 'We were impressed with how the IEA operated,' the unit director Digby Anderson explained. 'Just as it attacked a sclerotic consensus in economics, we wanted to do something similar on social policy.'[46] The publications usually select an issue of concern, such as relations between police and racial minorities or the housing problems of Asians, and then report how the issue has been covered in academic literature. Often authors find that the received view in academe has a poor evidential base, or may even fly in the face of evidence. These publications are designed, written, edited, and promoted to make such academic debate available to a wider audience. Anderson's weekly column in *The Times* provides a platform for such views.

The flavour of some of its authors' standpoints is reflected in the title of such hard-hitting collections as *Breaking the Spell of the Welfare State* (by Digby Anderson, June Lait, and David Marsland), which advocated swingeing cuts in state welfare spending, or *Pied Pipers of Education*, which argued for vouchers and fixed-term contracts for teachers. Another collection, critical of public-sector housing, is entitled *Home Truths*. The unit's Research Reports have discussed police accountability, criminal deterrence, social security reform, and positive discrimination. A substantial collection of essays, *The Kindness that Kills: The Churches' Simplistic Response to Complex Social Issues* was deeply critical of the way Church reports have analysed social and economic issues. And in late 1985 the unit published a series of pamphlets on religion and poverty which sharply contradicted the sociological and theological assumptions of the Archbishop of Canterbury's Report on inner city areas that appeared at the same time. At the time of writing new publications are in progress on the family, the curriculum, and nutrition policy.

The work of the unit's authors has sounded many favourite New Right themes. One has been the questioning of social engineering on the grounds that the engineer, social reformer, or intervener knows less than he assumes about what he is doing and little about the effects his actions will have. The criticism is of 'a notion which seeks to alter the behaviour of individuals, the characteristics of social groups, and the relations between individuals and social groups'. In many areas of social work or health education, SAU authors observe that the reformers lack an assured knowledge base and in consequence have difficulty in evaluating the results of their programmes and justifying public expenditure on them. Apologists for the programmes usually point to the need for yet more inputs of resources (of money, staff, and research), more time, and a redefinition of the programme's goals; or they offer an explanation of failure by pointing to other factors which are beyond the control of policy-makers. Anderson in particular remarks on the unassessable, unstoppable, irreducible character of state welfare—a system which he argues is literally out of control.

Some unit authors have also criticized the poor quality of much social science as it bears on social policy, pointing to studies, for example in poverty, welfare, industrial relations, which are shot through with relativism and value judgements and which are vitiated by ignorance about cause–effect relationships. The argument that the study of society is not and cannot be truly scientific is familiar. The material—human behaviour—is very different from the inanimate objects dealt with by hard science. People may change their behaviour, not least in the light of 'newly-discovered' information. It is not that the research findings in the social sciences are incorrect, but that their application in policy should be modest and tentative. Another theme in unit publications has been the argument that just as private industry producers have vested interests, so do producers of nationalized welfare and education. They deploy claims about 'service' and 'clients' and 'professionalism' rhetorically to manipulate public funds.

The unit is not a primary research institute or a policy centre. According to Anderson, 'the solution is not in politicians, but in a better informed public opinion'. This is done by making ideas available through its publications, through its authors' appearances in the media, and through media discussions of its publications so that ideas originating in academic research are successively translated into common currency. The unit, like the IEA is also useful in providing a rallying ground for like-minded people.

Adam Smith Institute

The Adam Smith Institute (ASI) was established in 1977 to develop free-market policies. The task of the ASI, according to its charter, is to further 'the advancement of learning by research and public policy options, economic and political science, and the publications of such research'. Hayek is chairman of its advisory board and it is run by two St Andrews University graduates, Madsen Pirie and Eamonn Butler. A number of other St Andrews graduates who became Conservative MPs, like Christopher Chope, Michael Forsyth, Robert Jones, and Michael Fallon, have done papers for the institute. The ASI is fairly libertarian on personal and economic issues and is deeply critical of public-sector economics. Pirie describes the public sector as 'inherently evil'.

The institute's work is distinguished by a concern with policy implementation, support of the free market, and the wide range of its proposals. It has produced a number of studies on quangoes, providing an exhaustive listing of such bodies, their budgets and terms of reference, and monitors the government's record in abolishing them. It has promoted policies of privatization, deregulation, and the contracting-out of services from local government. It has advocated the replacement of domestic rates, using asset sales to achieve cuts in the basic rate of income tax, and has even proposed a scheme for the private management of prisons. In *Inspector at the Door*, it enumerated the many powers of access to private property enjoyed by public officials and called for their reduction.

The institute has been particularly active in promoting privatization across a wide range of services including the nationalized industries, pensions, and local authority services. It has also argued for tax rebates rather than vouchers for persons who pay for children to attend private schools. It proposed the introduction of freeports into Britain. The idea of freeports was originally floated in the American magazine

National Review (January 1981). A survey was commissioned by the then trade minister, Ian Sproat, which was completed in December 1981 and recommended that freeports be set up near airports (the advantage of such a location was that the land was owned by the Department of Trade and Industry and could therefore bypass Treasury objections). These areas were established in 1983, are treated as outside the customs area, and goods may be manufactured, processed, and stored without payment of customs duty, and then exported to other countries.

The institute is interested in promoting public debate and awareness of its ideas. A good example of its approach was seen in an article on privatization in the short-lived magazine *Now*, on 13 June 1980. This was xeroxed by members of Mrs Thatcher's Policy Unit and sent to local Conservative councillors. The institute followed this by mailing 25,000 information packs to local policy-makers. But it is more concerned with practical policy than with propaganda. According to Pirie, 'a concrete example of action is the most effective form of propaganda'. The institute's main energies in the period 1982–5 were spent on the so-called Omega Project, a review of the complete range of government activity. Teams consisting of people from business, the media, public affairs, and the universities were assigned to examine the work of each department. The remit of each group was to suggest what activities could be transferred from the public to the private sector and where opportunity for choice and enterprise could be opened up. The conclusions of the working parties were first published singly, then presented in a bulky publication, *The Omega File*, in 1985. The teams avoided direct contact with departmental officials and only examined documents which were in the public realm. It proposed a massive programme of privatization of local and central government services and industries. Private insurance should replace the comprehensive welfare state and in the National Health Service Regional Health Authorities should be abolished and more use made of encouraging private health insurance, charges, and contracting out. In education parents should be given choice over schools and, via school boards, they should determine a school's policy.

Centre for Policy Studies

The Centre for Policy Studies (CPS) was established by Sir Keith Joseph in August 1974, shortly after the Conservative party's defeat in the February election. The original inspiration was Sir Keith's desire to learn the lessons from the Conservatives' experience of government between 1970 and 1974 and to study the successful social market economies, like Japan and West Germany. It was established with the permission of the leader, Edward Heath. Sir Keith Joseph became chairman, Margaret Thatcher president, and Alfred Sherman (later Sir) became the first director. The Conservative Research Department of course reflected the policy outlook of the leadership, and it was this that Sir Keith wished to question. Challenging established views and widening the range of free-market options for a future Conservative government was, it was argued by Sir Keith's backers, best done by an organization which had some distance from the party machine.

The political importance of the CPS for internal Conservative politics became apparent in September 1974 when Sir Keith openly broke with the Conservative leadership on economic policy. In his famous Preston speech in that month he emphasized the important role of the money supply as well as incomes policy in any attack on inflation (see below). Sir Keith's speech was widely viewed as a bid for the leadership, even though it was Mrs Thatcher who later toppled Mr Heath. Because the Conservative Research Department at the time still reflected the influence of the ideas of R. A. Butler and One-Nation Conservatism (see below p. 190), Mrs Thatcher turned to the CPS for support and it became an ideological rival. But its importance stemmed largely from the work done by CPS-connected people acting in their own capacity and in advising ministers and Mrs Thatcher. For example, among its early associates were Alan Walters (later Sir) and John Hoskyns (later Sir), who were to work for Mrs Thatcher at 10 Downing Street. Hugh Thomas (later Lord Thomas of Swynnerton who became chairman in 1979) and Alfred Sherman were active in speech-writing for

Mrs Thatcher, and Thomas has also chaired a Conservative policy-making group on the EEC.

The centre has operated by helping to publish the speeches of Sir Keith Joseph and Margaret Thatcher, presenting evidence to parliamentary and select committees and establishing study groups to examine particular measures. The reports favour market-orientated policies and 'should deal with the "how" rather than the "why" issues of policy', according to Elizabeth Cottrell, its Director of Research until 1984. Such study groups on British Telecom, British Steel, and British Leyland have pursued a generally anti-statist line. The latter resulted in the pamphlet, *British Leyland: A Viable Future* (1981) and advocated the privatization of profitable parts, and in due course this became government policy. It also advocated the abolition of the Inner London Education Authority. It is critical of many aspects of comprehensive education and published a study of examination results by pupils in different types of secondary schools, *The Right to Learn* by Baroness Cox and John Marks. Its transport group had advocated the conversion of infrequently used railway lines into roads. Other published reports have advocated portable pensions, the prohibition of strikes in essential services, and giving parents greater power on the governing bodies of schools. Among its more notable publications have been *Class on the Brain*, by P. T. Bauer, *Second Thoughts on Full Employment*, by Sam Brittan, and *Monetarism: An Essay in Definition*, by Tim Congdon. By the end of 1985 the CPS had produced over eighty substantial publications.

A number of influential figures around Mrs Thatcher have been associated with the CPS. These include people like Sir John Hoskyns, Lord Thomas of Swynnerton, Sir Alan Walters, Lord Young, and Sir Alfred Sherman. Sherman's original idea was that the centre should be distinct from the Conservative Research Department, to provide a forum for the exchange and development of ideas, cultivate public opinion, challenge prevailing assumptions, and advise ministers. He wanted the centre to retain what he regarded as its original ginger-group function and not become too tied to the Conservative Party, retaining the capacity to criticize the Conservative government from the free-market point of view and attract other supporters and thinkers who were not necessarily pro-Conservative. In a radio interview, Sherman commented: 'I saw it in the first place as an out-rider. An organization that would not be in the Party and therefore would be able to ask questions, to think the unthinkable, to question the unquestioned . . .'.[47] In a later lecture at the London School of Economics on 2 May 1984 he claimed that he did not want 'to be drawn into competition in courtship and sycophancy'. Sherman finally left the centre and established a new research body in early 1985. Lord Thomas had a more modest view of the centre's role and thought any criticism which its publications made of the government should be measured and constructive. The centre is bound to be somewhat inhibited in expressing criticism of the government as long as its president is the Prime Minister. But its *Whither Monetarism?* in December 1985 contained the Chancellor's Mansion House speech as well as critical comments on economic policy from three economists, Patrick Minford, Lord Bruce-Gardyne, and Tim Congdon.

Other Groups

There exist also a number of interest groups which are not primarily concerned with influencing public opinion but do promote general ideas and policies that are broadly similar in inspiration to those of the groups discussed earlier. We stress that these are interest groups, because political propaganda and education campaigns are only one of their aims. One long-standing body is Aims of Industry, which was founded in 1942 as a lobby for private enterprise. It became famous for its hard-hitting, anti-nationalization advertising in the late 1940s and the 1950s. Its general policy stance, advanced in vigorous press campaigns, is critical of existing levels of public spending and direct taxation and of government controls and regulations over industry on the grounds that these stifle initiative. More recently it has campaigned for tougher controls on trade unions and the abolition of the Greater London Council.

The Institute of Directors was formed in 1906. At present it has some 30,000 members, most of whom are chairmen or managing directors of companies. It emerged as a more vigorous spokesman for free enterprise under Walter Goldsmith, its director-general between 1980 and 1983. He proved to be a skilful publicist and a committed supporter of Mrs Thatcher on cutting public spending, tackling the trade unions, and rejecting the 'social partners' approach associated with the more 'official' CBI (Confederation of British Industry). The institute has differed from the latter body in being wholeheartedly in favour of the government's privatization programme. The CBI's ability to speak boldly for free enterprise is limited by its close involvement in such corporate bodies as NEDC (National Economic Development Council) and the Advisory Conciliation and Arbitration Service (ACAS) and the inclusion of nationalized industries among its membership. There was, therefore, an opening in the 1970s for a more outspoken defender of free enterprise of ideas and Goldsmith took advantage of it. Although it is not a think-tank it has a Policy Unit of six members (which is larger than that of the CPS or IEA) who work on such matters. It has managed to establish regular contacts with senior civil servants, political advisers in Whitehall, and the Policy Unit in 10 Downing Street. The institute's support for lower taxation and a reduction of public spending was reflected in its submission to the government's review of benefits for children and young people in July 1984. It favoured a cut in child benefits, a shift to child tax allowances, a major reduction in state provision of welfare benefits, and a welfare or social security tax separate from income tax. It has also advocated the concentration of benefits on the needy and more means testing of benefits and argued that any reform of the benefit system should be sensitive to the effects of welfare on the economy. The IOD's interest in free enterprise and its hostility to public ownership leave little doubt about which party it prefers at the end of the day.

The mid-1970s also saw an upsurge of predominantly middle-class (or *petit bourgeois*) protest groups reacting to the collectivist–corporatist approach of both Tory and Labour governments.[48] Aggressive ratepayers' groups sprang up in 1974–5, a year of heavy rate increases. Groups of small businessmen and the self-employed (for instance, the National Association of Self-Employed) complained of being 'squeezed' by high taxes, high rates, high inflation, and the growing influence of the TUC and CBI. Particular targets for their wrath were the Employment Protection Act (1975), which made it difficult and/or expensive to dismiss workers, and the Social Security (Amendment) Act (1975), which levied an extra 8 per cent national insurance contribution on the self-employed. Although the leaders of these groups could complain about the Conservative leadership it was largely that party, rather than Labour, to which they looked for support.

The National Association for Freedom was founded in December 1975 in the wake of the murder of Ross McWhirter by the Irish Republican Army. It shares with the economic groups a wish to limit the influence of the state. But in addition to promoting private enterprise it has been interested in the protection of personal liberties (particularly against trade unions and local authorities) as well as the introduction of a Bill of Rights and a Supreme Court in Britain. It also favours a strong military defence to protect the British way of life against the designs of the Soviet Union. The Association publishes a newspaper *The Free Nation* and has established itself as a successful campaigning body. It is not formally attached to the Conservative party but has a number of Tory MPs on its National Council. Although it wishes to go beyond the official party policy in some areas—for example, it favours the complete abolition of the closed shop—Mrs Thatcher is an admirer of its policies. It has successfully used the courts on several occasions against trade unions. It was active in providing legal advice for George Ward, the owner of the Grunwick factory, in his bitter dispute with striking members of the APEX union over recognition of the union's negotiating rights during 1976 and 1977 and provided legal advice for the Conservative controlled Tameside Education Authority in its dispute with the DES about selection of children for secondary education in 1976. It also took out injunctions against the Post Office in 1977, when postal workers refused to sort mail destined for South Africa, and took British Rail to the European Court of Human Rights in 1979 over the dismissal of three of its employees on the grounds that their non-membership of a rail union offended its closed shop policy. In 1981 it also promoted the appeal of Joanna Harris, a local authority poultry inspector, against her dismissal for refusing to join a union.

Within the Conservative party there are two well-established groups which promote right-wing views. The Monday Club was formed in 1961 in protest at the then Conservative government's policy of a speedy move to self-government for British colonies in Africa. In 1985 it had eighteen MPs and fifteen peers as members, though none was in the Cabinet. The group pursues a free-market line on economic and welfare issues and supports a 'hard' line on issues like immigration, race-relations, and capital punishment.[49] It attacks the idea that Britain is a multiracial society and favours the abolition of the Commission for Racial Equality. The Selsdon Group was formed in 1973 to protest at the collectivist drift of the Heath government, particularly its rescue of the financially troubled firms of Rolls–Royce and Upper Clyde Shipbuilders. It was named after the Selsdon Park Hotel, where the party discussed its manifesto plans for the 1970 general election. It supports free-market policies, privatization of nationalized industries, education vouchers, private provision in health and welfare, and major cuts in taxation and public spending.[50] In 1985 the group had eight MPs and one member, Nicholas Ridley, in the Cabinet. Another Selsdon member is Ian Gow, Parliamentary Private Secretary to Mrs Thatcher between 1979 and 1983 and then a junior minister until November 1985.

Permeation

The impact of these ideas on the Conservative party under Mrs Thatcher was helped by mass media coverage. In the press, *The Economist*, *Daily Telegraph*, *Spectator*, and *The Times* (after Charles Douglas-Home succeeded Harold Evans as editor in 1980) increasingly provided editorial support for many of these ideas. The columns of the quality newspapers were regularly opened to writers like Paul Johnson, Roger Scruton, John Hoskyns, Digby Anderson, and Walter Goldsmith. The tabloids, though less interested in politics, have been generally hostile to high taxation, abuses of trade union power, and 'welfare scroungers'. The *Sun* had a regular column from John Vincent, a history professor at Bristol University and the *Express* one from George Gale. A number of commentators like Woodrow Wyatt, Hugh Thomas the historian, John Vaizey the economist, and Paul Johnson, ex-editor of the left *New Statesman*, as well as former Labour Cabinet ministers like Lord George-Brown, Lord Richard Marsh, and Reg Prentice, moved to the right in the 1970s. The political imbalance of the daily press in Britain is well known and has long been hostile to the left of the Labour party; in 1983, some three-quarters of the British population was reading a daily paper advocating a Conservative victory. The left of centre contributors to the *Guardian* and the *New Statesman* have hardly competed with the agenda-setting role of the New Right publicists. In so far as there is a core of support for the new thinking it is probably best found in the columns of the *Daily Telegraph* and *Spectator* and among the members of the working groups at the CPS and regular lunches at the IEA. A rather different, more traditionalist set of right-wing values is propounded by the monthly *Salisbury Review*, founded in 1982 (see below p. 105).

None of the groups discussed in this chapter have a formal connection with the Conservative party, although individual members are involved in the work of the party. A number, for example, worked on the policy groups which operated in 1982–3 in the preparation of the party's election manifesto for 1983. John Wood of the IEA was a member of the Why Work? committee, Lord Thomas of the CPS chaired the group dealing with Europe, and Walter Goldsmith of the Institute of Directors was on the enterprise and privatization group. Since the 1983 general election various groups have been considering possible initiatives in policy areas. Lord Thomas of Swynnerton chairs a group on the British government's relations with other states, Baroness Cox of the CPS chairs one on education, and Nigel Vinson one on personal capital. Until 1985 the work of the groups was co-ordinated by Elizabeth Cottrell, then Research Director of the CPS.

The groups and individuals represent different aspects of New Right issues. But the personal and institutional links between them constitute a network. Their personal contacts with Mrs Thatcher or her Policy Unit or others to whom she and ministers may listen and their appointments to official bodies and access to the media are methods of permeation. They have not dictated the government's policies; they have to compete with rival lobbies and the opposition of some sceptical ministers and senior civil servants. But they have been a clear leader of radical policy reviews such as the Fowler review of social security (1983–5) and policies such as privatization and education vouchers; they have sternly criticized back-sliding ministers; and, through their suggestions, they have shaped the agenda of policy discussion. A more direct 'input' of New Right ideas was made by Sir Keith Joseph, when he became Secretary of State for Industry in 1979. He issued a 'reading list' to his senior civil servants which was dominated by conservative and free-market writers, including Hayek, Friedman, and von Mises.

The groups are free to criticize the Conservative leaders, and their contributions to public debate and to the policy agenda can be exploited by any political party. There have been inevitable tensions between what may be called the intellectuals, who are primarily interested in ideas, and the politicians, who have responsibility for acting on them. The former are interested in promoting new ideas to change the climate of opinion and expand the parameters of policy discussion; the latter are interested in policies which are politically and administratively acceptable and electorally popular. A number of writers about modern Conservatism have also been involved in occasional seminars of the Conservative Philosophy Group, which are sometimes attended by Mrs Thatcher. Some are MPs, some are academics like Roger Scruton, Maurice Cowling, John Casey, and Edward Norman (all with a past or present connection with Peterhouse College, Cambridge, and all frequent contributors to the *Salisbury Review*) and others like Sir Alfred Sherman, Lord Hugh Thomas, Paul Johnson, and T. E. Utley are authors and journalists.

When one talks to members of the groups one is struck first of all by a pervasive mixture of optimism. This stems from the feeling that the tide is turning in their favour after a long period in which many of their ideas and policies were regarded as 'crackpot', even among Conservatives. Arthur

Seldon justified his optimism not on the grounds that politicians in all main parties were more appreciative of the role of markets than a decade earlier, but that social and technical changes were working that way. The *embourgeoisement* of the working class would encourage the further spread of home-ownership, private health arrangements, and private pensions. The growth of the 'black economy' was seen as another sign that people were turning away from government. In an editorial in *Economic Affairs*, October–November 1985, he wrote:

> The more the market is re-introduced by this Government, the less any Government of the future will be able or will want to suppress it. Whichever Party or coalition is elected to office in 1987–88 or 1991–92, the market is less likely than at any other time in the last 10 years to be removed where it has been restored or introduced, or to be repressed where social advance and technological innovation are forcing it to the fore. In time all British Parties will have to accept the market not only because it is more efficient than government but no less because to deny it is to forfeit political power.

Yet there is also a sense that the present conjunction of circumstances—the electoral collapse of Labour in 1979 and 1983, weakness of the trade unions and Mrs Thatcher's dominance in the Conservative Party—provides the most favourable conditions for the implementation of these views and may not last much longer. It is a case of 'now or never'.

The admiration for Mrs Thatcher and the awareness that she is the most sympathetic party leader for these ideas is tempered by a scepticism about all politicians, who are widely regarded as opportunists. Doubts are frequently expressed about the views of some of her senior colleagues, in particular Peter Walker, John Biffen, and those who urge 'caution' or 'consolidation' rather than a further instalment of tax-cutting and free-market policies. Right-wing critics point to the levels of taxation and public spending which, as a proportion of GDP, are now higher than when the Conservatives came into office in 1979. They point to the lack of action on rent control, school vouchers, and student loans in higher education, failures to make trade union members contract in rather than contract out of the political levy, unwillingness to be more radical in curbing the increasing costs of the welfare state and

National Health Service, and the frustration of plans to end such middle class 'perks' as tax relief on mortgage interest payments and private pension contributions. The present government, they claim, has bowed to powerful lobbies, the caution of the civil service, and electoral expediency. They can also point to occasions when Mrs Thatcher herself has been seen to temporize. During the 1983 general election she frequently boasted about how much the government was spending on welfare and on jobs, and made the famous claim that 'the National Health Service is safe with us'. In January 1984 in an interview in *Weekend World*, she claimed that the existing level of public expenditure was 'about right', the remarks of a consolidator rather than a radical reformer.

Conclusion

Some indication of the distance between the government and the groups is seen in the regular budget recommendations from groups like the Institute of Directors and the Adam Smith Institute. In discussions preceding his 1984 budget, Nigel Lawson bluntly dismissed the practicability of the ID's suggestions on tax cuts. In 1985, for instance, the Adam Smith Institute proposed swingeing cuts in direct and indirect taxation, abolition of the higher levels of income tax, reduction of the standard rate from 30 to 20 per cent, abolition of capital taxes, and the introduction of tax relief for expenditure on private education and health, all to be financed by sales of state assets. But the government turned a deaf ear to the demands. Aims of Industry has campaigned for the abolition of wages councils; in 1985 the government acted but only to wind up councils covering workers under 21. The institute failed to get postal ballots for the election of union leaders and contracting in for the political levy inserted into the 1984 Trade Union Act. As Education Secretary Sir Keith Joseph took no action on loans for students and announced that the government had abandoned its intentions to introduce education vouchers. In turn, the scheme's supporters claimed that the civil service had been obstructive.[51]

The next chapter examines whether the ideas of these groups actually amount to a New Right. An affirmative answer has to demonstrate both an internal coherence and a differentiation from the Old Right. There is always a problem in labelling a set of ideas proposed by different people for different issues in different contexts. The danger in labelling, or typing, is that one loses an awareness of the subtle differences or nuances between groups. For example, Friedman and Hayek differ over the responsibility of trade unions for inflation; the Adam Smith Institute and the IEA differ over education vouchers. The groups and writers reviewed in this chapter represent various schools of thought, including Chicago economics (Friedman), classical liberalism (Hayek), and libertarianism.

Many of the groups have been suspicious of the Conservative government and are more accurately regarded as supporters of Mrs Thatcher and her views than of the Cabinet or the Conservative party as a whole. Another way to characterize the groups politically is by what they are not: they certainly are not pro-Labour, pro-Alliance, nor pro-One Nation Conservative.

Notes

1. (London, Longman, 1963), p. ix.
2. Ibid., p. xiii.
3. *The General Theory of Employment Interest and Money* (London, Macmillan, 1936), pp. 383–4.
4. *Disraelian Conservatism and Social Reform* (London, Routledge & Kegan Paul, 1967), p. 212.
5. *Our Partnership* (London, Longman, 1948), pp. 67–8.
6. A. M. MacBriar, *Fabian Socialism and English Politics 1884–1914* (Cambridge University Press, 1962).
7. D. Donnison, *The Politics of Poverty* (Oxford, Martin Robertson, 1982).
8. L. J. Sharpe, 'The Social Scientist and Policy Making: Some Cautionary Thoughts and Transatlantic Reflections', *Policy and Politics* (1975).
9. 'Policy-Making and Policy-Makers', in R. Rose (ed.), *Policy-Making in Britain* (London, Macmillan, 1969), p. 367.
10. *Human Nature in Politics* (London, Constable, 1948), p. 83.
11. S. Finer, 'The Transmission of Benthamite Ideas', in G. Sutherland (ed.), *Studies in the Growth of Nineteenth-Century Government* (London, Allen & Unwin, 1972), p. 13.

12. *British Cabinet Ministers* (London, Allen & Unwin, 1974).
13. For example, see S. Brittan, 'The Irregulars', in R. Rose (ed.), *Policy-Making*, and J. Hoskyns, 'Conservatism is not Enough', *Political Quarterly* (1984).
14. *Poverty and Politics and Policy* (London, Macmillan, 1979), p. 5.
15. Ibid., p. 7.
16. See R. Rose, 'Disciplined Research and Undisciplined Problems', in *International Social Science Journal* (1976).
17. Q. Hogg, *The Case for Conservatism* (London, Penguin, 1959), p. 16.
18. On Powell's ideas see K. Phillips, 'The Nature of Powellism', in N. Nugent and R. King (eds.), *The British Right* (Farnborough, Saxon House, 1977); P. Utley, *Enoch Powell: The Man and his Thinking* (London, Kimber, 1968); and D. Schoen, *Enoch Powell and the Powellites* (London, Macmillan, 1977).
19. J. Wood (ed.), *A Nation Not Afraid: The Thinking of Enoch Powell* (London, Batsford, 1965), p. 15.
20. D. Schoen, *Enoch Powell and the Powellites*.
21. (London, Unwin, 1943.)
22. *Two Cheers for Capitalism* (New York, Basic Books, 1978).
23. See E. Norman, 'Established Thinking as a Threat to Capitalism', *Times Higher Educational Supplement*, 13 May 1977.
24. (London, Heinemann, 1974.)
25. A. Halsey and M. Trow, *The British Academics* (London, Faber, 1971).
26. J. Tunstall, *The Westminster Lobby Correspondents* (London, Routledge and Kegan Paul, 1970).
27. C. Hewitt, 'Policy Making in Post-War Britain', *British Journal of Political Science* (1974).
28. For such a claim see R. Boyson, *Centre Forward* (London, Temple Smith, 1978).
29. J. Wood, *Powell and the 1970 Election* (Kingswood, Surrey, Elliot Right Way Books, 1970), p. 108.
30. R. Blake, *The Conservative Party From Peel to Thatcher* (London, Fontana, 1985).
31. B. Pimlott, *Fabian Essays in Socialist Thought* (London, Heinemann, 1985).
32. L. Hartz, *The Liberal Tradition in America* (New York, Harcourt, Brace, and World, 1955).
33. (London, Routledge & Kegan Paul, 1944.)
34. (London, Routledge & Kegan Paul, 1960.)
35. (London, Routledge & Kegan Paul, 1973, 1975, 1979.)
36. *The Constitution of Liberty*, p. 398.

37. S. Brittan, 'Hayek, The New Right and the Crisis of Social Democracy', *Encounter* (1980), p. 35.
38. *The Constitution of Liberty*, p. 63.
39. *A Monetary History of the United States 1867–1960* (NBER, Princeton University Press, 1963).
40. Interview.
41. W. H. Greenleaf, *The British Political Tradition* (London, Methuen, 1983), Vol. ii, p. 33.
42. *Over-Ruled on Welfare* (London, IEA, 1979).
43. *The Emerging Consensus* (London, IEA, 1981), p. 105.
44. *After the New Right* (London, Heinemann, 1983), p. 82.
45. Interview.
46. Interview.
47. Interviewed by Peter Hennessy on BBC Radio 3, 21 October 1983, 'The Think Tank'.
48. R. King and N. Nugent, *Respectable Rebels* (London, Hodder and Stoughton, 1979).
49. P. Seyd, 'Factionalism within the Conservative Party; The Monday Club', *Government Oppostion* (1972).
50. M. Loveday (ed.), *Parliamentary Index* (London, Adam Smith Institute, 1982).
51. M. Seldon (ed.), *The Riddle of the Voucher* (London, IEA, 1986).

4

A NEW RIGHT?

A NUMBER of commentators have discerned a convergence in the ideas of the groups discussed in Chapter 3 but disagree about calling them 'right' or 'liberal'. Are the ideas old fashioned liberalism or genuine conservatism or a blend of both? In his stimulating study of the social and economic doctrines of the New Right, Nicholas Bosanquet organizes the ideas around a *thesis* which sees society as having an inherent tendency towards order and justice that is reflected in the free market and an *antithesis* which sees this particular order of society as being damaged and distorted by the effects of government intervention.[1] The interventions usually result in increases in public spending and taxation and a loss of economic efficiency. The benign effects of the free market are counterposed to the defects of the politicization and centralization of decision-making. Bosanquet claims that these themes constitute a 'New Right' and in Britain they are identified with the Conservative party.

An alternative view is expressed by Norman Barry who rejects the New Right label. Barry notes that many of the arguments in what he calls 'The New Liberalism'—advocating a greater role for markets and individuals and a lesser one for the government—fall into two categories.[2] The 'rights' theorists, who defend the above policies primarily because they promote individual self-development, differ from 'consequentialists', who defend the policies because they promote desired social goals such as economic growth, efficiency, and so on. Libertarian philosophers, who are almost anarchist in their individualism, dominate the 'rights' theorists and economists dominate the 'consequentialists'.[3] This chapter discusses the problems involved in using the label New Right to characterize the groups. It then examines the core ideas of the groups and discusses how they have been promoted.

Conservative Liberalism

Bosanquet's perception of a coherence in the ideas probably derives from his focus on a narrow band of 'New Right' writers and theories, largely dealing with economics. Barry is more aware of work in other disciplines, particularly law and philosophy, and the ideas of American libertarians. An excellent analysis by John Burton also prefers to characterize the ideas as a 'new liberalism' rather than as a 'New Right'.[4] He notes that the ideas are variegated and cover three areas; liberal solutions to particular social and economic problems, developments in *positive* economics, particularly in microeconomics, and libertarian values. The inspiration for much of the thinking stems from Adam Smith, the Austrian School of Economics, and public choice theory, rather than Burke and Disraeli.

There are problems of terminology here and it is worth reflecting on the difficulties which attend the use of some of the labels for these movements both in Britain and the United States. There is a tendency to distinguish between the political *left* and *right* in referring to distinctive sets of political attitudes and interchange these with, respectively, *liberalism* and *conservatism* in the United States, and *socialism* and *conservatism* in Britain. Recently it has also become fashionable to talk, particularly in the United States, of a 'new' left or right and of *neo*-conservatives or *neo*-liberals. In the United States many liberals, since the period of the New Deal in the 1930s, have favoured the intervention of the federal government to promote welfare, civil rights, and employment. They have also favoured an interventionist role for the US in international affairs. American Conservatives, by contrast, have traditionally wanted to limit the role of federal government and have inclined to an 'isolationist' stance in foreign affairs. In more recent years a group of intellectuals (including Irving Kristol, Nathan Glazer, and Daniel Moynihan)—who in the 1960s had been liberals and had spoken out for active government—became disillusioned with the consequences of the programme and promoted a neo-conservative backlash.[5] Many old liberals are now conservative. A problem with using the term 'liberal' in Britain is that in the closing decades of the nineteenth century liberalism itself was dividing into 'new' and 'old' versions. Old or classical liberals had taken a negative view of the state, regarding most of its interventions as an infringement of human liberty and choice. Freedom was negative and it arose from the absence of state control. Towards the end of the century T. H. Green and others took a more favourable view of the state's ability to promote individual freedom (for example, by means of compulsory education). The state could 'positively' assist freedom. This gave rise to a school of 'New Liberals' who, in the first two decades of the twentieth century, accepted a greater degree of collectivism. Indeed some eventually moved over to the Labour party, regarding it as an effective vehicle for progressive liberalism.[6] I shall argue that in spite of the clear tensions between liberal ideas about economic policy and non-liberal, even authoritarian ideas in other areas (see below), the term 'New Right' is acceptable as a description of the British case. Compared to the United States, the libertarian ideology has been much less widespread in Britain.

There is also some difficulty in using the terms left and right, as if all issues can be located on a single left–right dimension. The assumption is that a person's position on one issue entails a predictable position on another issue, for instance a supporter of nationalization is against nuclear weapons. In common usage the left–right dimension is usually applied to economic issues, with Labour (advocating a major role for the state in the economy) placed on the left and the Conservatives (preferring free enterprise) on the right, and similar labels are attached to policy groups in the two parties. However, Sam Brittan has fairly objected to the use of one dimension for all issues.[7] Conservatives may be generally opposed to state intervention on economic matters but favour state intervention on social and industrial relations issues, being right-wing on one set of issues, left on another. For example, Enoch Powell's liberalism embraces both free-market economics and social issues (for example, he regularly votes against the restoration of capital punishment and in 1965 was co-sponsor of a parliamentary bill to legalize homosexuality between consenting adults). But he is distinctly 'illiberal' about immigration. Most Labour MPs favour state intervention in the economy but are more *laissez-faire* on industrial relations and many social issues. Barry's 'New Liberals' are consistent 'liberals', who emphasize the importance of personal freedom and limited government in both the economic and non-economic spheres. But many British Conservatives combine liberalism on economic issues with authoritarianism on other issues.

One source of unease about the label 'New Right' among many of its members is the association of the political right with authoritarianism on many social issues.[8] This tension clearly emerges in a recent book of essays by young writers entitled *The 'New Right' Enlightenment.*[9] Although most of the essayists support the Conservative party, their libertarianism usually makes them only qualified supporters and they are concerned about many acts of political centralization by the Thatcher government. In Britain in the 1980s there are two distinct groups who may claim to represent the 'New Right'. One is found among the proponents of the free market described in Chapter 3. The other centres on the academics and journalists who are associated with the *Salisbury Review.* Their outlook is well expressed in Roger Scruton's *The Meaning of Conservatism*, which rejects the free market and defends the authority of the state and 'a view of legitimacy that places public before private, society before individual, privileges before right'.[10] This view is very different from the libertarian, free-market, and, in some cases, anarchist approach of the *Enlightenment* essayists. Roger Scruton and other members of a Peterhouse School of Conservatism (because a number of them, like the academics Maurice Cowling, John Casey, and Edward Norman, and journalists like George Gale, Colin Welch, and Peregrine Worsthorne have a past or present connection with the Cambridge college) are more worried by an alleged excess of democracy and permissiveness. The views were boldly expressed in contributions to *Conservative Essays*, edited by Cowling in 1978. Worsthorne, for example, in an essay revealingly entitled '*Too Much Freedom*', called for the state, 'to reassert its authority, and it is useless to imagine that this will be helped by some libertarian mish-mash drawn from the writings of Adam Smith, John Stuart Mill and the warmed-up milk of nineteenth century liberalism'.[11]

The *Essays* and *Salisbury Review* combine social authoritarianism with an interest in race, nationhood, national identity, and patriotism. Not surprisingly Enoch Powell is something of a hero for this group. It is also among these writers that one finds support for the nearest equivalent to the moral majoritarianism of the United States. Conservative critics of alleged 'permissiveness', lawlessness, weakening authority of teachers, parents, employers, and police, and decline of self-restraint have been the most conspicuous defenders of various 'moral' cause groups in the 1980s. These include Mary Whitehouse's National and Viewers Association, which campaigns about bad language, violence, and sex on television; the Festival of Light; and anti-abortion groups like Life and the Society for the Protection of Unborn Children (SPUC). They have been the main parliamentary supporters in 1986 of Winston Churchill's defeated bill to bring television programmes within the scope of the Obscene Publications Act which bans material tending to deprave or corrupt, and in 1985 for Enoch Powell's private members' Unborn Children (Embryo) Protection Bill, which sought to prevent embryos being created or retained for any other purpose than enabling a child to be born. They have also been supportive of Mrs Victoria Gillick's campaign to refuse doctors permission to prescribe contraceptives to girls under sixteen without parental consent. Responsibility for this moral decline is fastened on the leaders of the counter-culture, particularly progressive teachers, social workers, lenient magistrates, feminists, and race relations advisers. This wing of the New Right is concerned above all with social order.

It has been writers of the political left who have been most sensitive to the 'gap' between the New Right liberalism on economic issues and authoritarianism on social issues.[12] Moreover this governmental authoritarianism, in the guise of greater central government control, has been felt by many other groups—local government, trade unions, civil service and higher education (see below p. 286). 'Rolling back the state' is therefore selective. The fusion of the free-market (neo-liberal) and social authority (neo-Conservative) elements is clearly expressed in Mrs Thatcher's beliefs (Chapter 1) and record (Chapter 10). A brilliant attempt to distinguish the two has been made by Andrew Beesley, who claims that neo-liberalism is constructed from five points which, in order of importance, are: (1) the individual; (2) freedom of choice; (3) market security; (4) laissez-faire; (5) minimal government. Neo-Conservatism takes these five points in reverse order of importance. Its politics is therefore constructed from; (1) strong government; (2) social authoritarianism; (3) disciplined society; (4) hierarchy and subordination; (5) the nation. The political practice of Thatcherism combines both sides.[13]

Core Ideas

A review of the main ideas of the free-market groups does reveal a certain coherence. First and foremost is the general suspicion of government activity, which goes beyond the modest duties laid down by Adam Smith, of providing national defence, law, order, and justice, and certain public works. The government's strict control of the money supply is regarded as an essential tool in the attack on inflation. Burton has correctly pointed to different types of monetarists—'fiscal monetarists', who argue that control of the budget deficit is necessary for curbing money stock growth, and Friedmanites, who reject any necessary connection between the size of the budget deficit and monetary growth. But both agree that inflation is a monetary phenomenon and that it is politicians, surrendering to pressure, who are too lax in controlling the supply. The need, as it were, is to protect the economy from vote-seeking or group-appeasing politicians. There is reservation about the role of government as an economic manager, social planner, and provider of welfare. There is also general hostility to the public sector, nationalized industries, state provision of universal flat-rate welfare benefits as of right, and existing levels of direct taxation, subsidies, and public spending. Government action is widely regarded as having failed when it has replaced the market in delivering services, particularly in health and education. State action may also interfere with spontaneous acts of charity and self-help.

This outlook is tied to a suspicion of the political process which is too easily manipulated by self-seeking groups—

producer groups, lobbies, bureaucrats, and other special interests. Politicians cannot be trusted—to control the money supply or cut taxation. As electorates have increasingly turned to governments for solutions to problems, so popular expectations have been raised to unrealistic levels and government become overloaded. Subsidies and public spending grow because the interests who benefit are concentrated, well-organized, and influential, whereas the costs are dispersed among millions of less well-organized tax-payers. So many public services, it is claimed, have become less responsive to consumers and more beholden to employees.

More positively, there are proposals to place constitutional limits on what government (and parliamentary and electoral majorities) can do. Such critics would sympathize with one of J. A. Schumpeter's conditions for a successful democracy—namely, that the effective range of political decisions should not be pushed too far.[14]

A second target is the role of social justice as a criterion for policy. Bosanquet notes that for the New Right there is no legitimate agenda for distribution: spokesmen are particularly suspicious of attempts to broaden the goals of equal civil rights and equal opportunity into equality of outcome.[15] They are also hostile to definitions of poverty in relative terms, which means that the criterion of anti-poverty policy is changed from the elimination of absolute poverty to redistribution along more egalitarian lines.

There is, thirdly, a dislike of many features of British trade unionism. The existence of legal privileges, such as the closed shop and picketing, not granted to other voluntary groups is obviously offensive to libertarians, but there is also hostility among free-market supporters to monopoly national wage-bargaining and the negotiation of national wage scales, regardless of local conditions and employers' ability to pay. The restrictive practices of trade unions are also identified as a major cause of uncompetitiveness, unemployment (or lack of responsiveness in labour markets), and slow economic growth.

A less salient theme has been foreign policy and Britain's place in the world. Apart from some CPS publications and statements by NAFF, foreign and defence matters have largely been left to specialist groups (such as the Institute for European Defence and Strategic Studies and the Coalition for Peace Through Security). These have organized conferences and seminars and published papers about security issues, and generally propagated pro-Atlantic Alliance and anti-CND policies. A number of commentators also expressed disappointment with the results of East–West *détente* in the late 1970s and the USSR's human rights performance according to the Helsinki agreement in 1975. Critics were particularly critical of that country's harsh treatment of dissidents and reactions to applications from Jews for exit visas from the USSR. Such cases were regularly covered in lengthy articles by Bernard Levin in *The Times*. Mrs Thatcher frequently expressed her admiration for the exiled Solzhenitsyn and his warnings that the West should not relax its defences. Her original hostility to the Soviet regime and determination to pursue a more assertive defence policy was also probably encouraged by the election of Ronald Reagan to the presidency in the United States. In that country a central plank on the revival of neo-conservative ideas was that the decline of America's world leadership, in the aftermath of Vietnam, Watergate, and the weakened presidency, had given scope for Soviet expansion.

The civil service is also often singled out for criticism. The bureaucracy is exempt from market conditions of labour and, as seen by the Virginia School of Public Choice, is expansionist, distrustful of the market, and frequently manages to 'capture' even reforming ministers. Many supporters of Mrs Thatcher who have some experience of Whitehall—like Norman Strauss, Sir John Hoskyns, and Sir Alfred Sherman—have vigorously criticized senior civil servants because of their alleged attachment to the established collectivist consensus and their lack of commitment to Thatcherite ideas.

Finally, the groups approve the working of the free market, on the moral grounds that it provides opportunities for individual choice, development of self-reliance, and responsiveness, as individual consumers can indicate their preferences. The market is also defended on utilitarian grounds, namely, that it is the best way of promoting economic growth, experimentation, adaptability, and learning. The more that market policies are pursued—ending special subsidies, tax concessions, rent controls and introducing education vouchers—the more power is removed from politicians and civil servants and restored to the people.

It is fair to claim that the above ideas do amount to a coherent statement of policies. But are they new? Mrs Thatcher and her supporters have at times found it convenient to claim to be radical and innovative compared to previous Conservative governments. Similarly some 'wet' Conservatives and the opposition parties also have an interest in claiming that she has broken with the old 'reasonable' Tory party and consensus politics. Many left-wing commentators have also attributed a remarkable coherence and power to the ideas of the New Right. According to Andrew Gamble in *Marxism Today* in July 1983, 'the New Right has won the battle to shape the new consensus on how to respond to the recession'.

In the twentieth century British Conservatism has contained two strands, a neo-liberal, individualist strand and a Tory, or collectivist, one (see Chapter 7). There is a general agreement that the drift of Conservatism, particularly when in government from the 1920s until 1974, has clearly been collectivist. Of course this has often been qualified in practice and sometimes reluctant, but the record is clear enough if one looks at the 1930s, the 1950s, the 1960s, and 1972–4. Many of the speeches of Neville Chamberlain, Sir Anthony Eden, Harold Macmillan, and R. A. Butler contained defences of the positive role of central government. The Conservative distrust of the market has been reflected in the protectionism for industry and agriculture in the 1930s, regional aid in the 1950s, economic planning and incomes policies in the 1960s, the backbench resistance to the abolition of resale price maintenance in 1963, and, finally, the statutory controls on prices and incomes and government intervention in industry in the 1970s. For much of the past century the Conservative leadership has been respectful of the status quo and sceptical of ideologies or policy packages like monetarism, and has shared an organic view of society. Under Mrs Thatcher, however, the emphasis on the virtues of the market and individualism and the open hostility to many aspects of the public sector, civil service, and trade unions are breaks with the traditions of the party leadership. Although Thatcher grants an important role to government in matters of law and order she gives it only a modest one as a social and economic guide.

A New Right?

There are differences in the interests and campaigning styles of the groups. Aims of Industry is interested in issues affecting private business, the IEA in the public sector, and the CPS and ASI in policy implementation. Yet the close personal links existing between many activists in these groups have helped to develop what might be called 'a free enterprise solar system'. Obviously the Conservative party is a primary focus of much of the groups' lobbying and research. It is worth noting, however, that the party's Research Department, directed by Christopher Patten between 1974 and 1979, was not sympathetic to the new thinking. Relations were particularly tense with the Centre for Policy Studies which, some Research people felt, was a rival for the ear of the parliamentary leadership. The formal attachment of Mrs Thatcher (as president) and Sir Keith Joseph (as director) with the CPS is the most obvious party link. But many of the other groups have recruited sympathetic Conservative MPs on their advisory boards or as authors or members of study groups. Another device for linkage is the personal interconnections between groups. Sir Alfred Sherman worked as a speech-writer and adviser to Sir Keith Joseph in 1974, introducing John Hoskyns and Norman Strauss (then at Unilever) to each other and then to Sir Keith and Mrs Thatcher. Hoskyns joined the CPS and, with Strauss, went to Number 10 in 1979 as adviser to Mrs Thatcher.

Lord Beloff, Chairman of the Conservative Research Department, is on the board of the CPS. T. E. Utley, an assistant editor of the *Daily Telegraph*, is on a CPS study group and the SAU board. Emeritus Professor Anthony Flew is on the boards of the ASI and Freedom Association and has been associated with the CPS. Baroness Cox is a member of the SAU board and CPS groups. Graham Mather worked for the Freedom Association, before becoming Director of Research of the Institute of Directors in 1980. The same overlap of personnel is seen in the literature published by the groups. Digby Anderson has written or contributed to publications for the SAU, Freedom Association, and IEA. Seldon, author of many IEA pamphlets, has also written a paper on the National Health Service for the CPS. Madsen Pirie, a prolific writer for ASI, wrote a pamphlet on the privatization of the telephone service for Aims of Industry.

The offices of the groups are located within a few square miles of each other in Westminster, close to Parliament, Fleet Street, and the Conservative Party headquarters. The regular lunches at the IEA and the CPS study groups provide mutual social and intellectual support for participants. 'In the early days we reassured free marketeers that they were not alone,' said Arthur Seldon.

On occasions the groups have also operated as a counter civil service or as a ginger group pressing more radical policies on ministers. A particular target was Mr Prior, who had little support for his 'softly softly' approach to industrial relations reforms when he was Secretary of State for Employment 1979–82. Some ministers, including Patrick Jenkin (when at the DHSS), have also asked the policy activists to debate issues with their civil servants. Draft proposals for the privatization of British Telecom were drawn up by two academics. Professor Beesley of the London Business School wrote 'Liberalisation of the Use of British Telecommunications Network' in 1981 and Professor Littlechild of Birmingham wrote 'Regulation of British Telecom's "Profitability" ' in 1983; significantly, these were not written by civil servants in the Industry Department. Groups have also taken advantage since 1979 of departmental reviews of policies —such as the Serpell inquiry on British Rail or the Fowler reviews of portable pensions and the welfare state—to submit criticisms and proposals. In 1983, the press reported that monthly lunches by a so-called Argonauts Club, consisting of Sir Alfred Sherman, Sir John Hoskyns, Michael Ivens, and Walter Goldsmith, discussed new lines of policy for industry and the trade unions and were often joined by members of Mrs Thatcher's Policy Unit.

Events in the 'real' world, particularly the growing concern over inflation, the obvious shortcomings of incomes policies and successive governments' management of the economy, and the intellectual crisis of Keynesian economics, encouraged policy-makers to look for new ideas. The expected 'trade off' (Phillips curve) between inflation and unemployment was breaking down, as many western economies suffered rising levels of both in the 1970s. Ideas of monetarism and privatization gained a hearing among a growing number of commentators and policy-makers in the 1970s in Britain and elsewhere as a way of coping with genuine problems of rising public spending and inflation. Control of inflation was the top priority for governments. Hugo Young noted in the *Guardian* the revival of interest in the ideas of Hayek in the mid-1970s and the conjunction of ideas and circumstances: '. . . the man and his ideas finally met their moment. They collided with a party traumatized by defeat, guilty about its failure, and vengeful against a leadership which had distanced itself by a succession of U-turns from the policy set out in 1970.' (8 October 1984.) For much of the post-war period the system of fixed exchange rates had imposed monetary control of a sort on governments. If domestic money supply increased too rapidly then inflation, loss of competitiveness, and balance of trade deficits usually led to heavy selling of a country's currency and to pressures for the restoration of 'sound' policies. By 1972 however, the currencies of most major countries were floating and no longer tied to a fixed exchange rate. In the vacuum money-supply figures gained increased significance as an earnest of a country's determination to master inflation. If the signals were negative, holders of the currency were likely to sell and create an exchange crisis.

Another reason for the progress of the ideas was that, as Nicholas Bosanquet points out, the New Right was prepared to reach out and relate its ideas to a political agenda covering not only economic policy, but also welfare and industrial and social issues. In becoming more 'professional', academic economists also became more quantitative and less relevant to many broader issues of social policy. But if the right had dominated much of the so-called 'new economics'—monetarism, public choice, and the revived Austrian school of economics—it still faced resistance from a Keynesian-dominated economic profession in Britain. Indeed monetarism gained more adherents among policy-makers than theoreticians. John Burton has described most British academic economics as 'the apologists of interventionism'.[16] In a letter to *The Times* in 1981, 364 economists, including four former chief economic advisers to government, stated that the government's policy of decelerating the rate of growth of money supply would worsen the depression and that 'there is no evidence in economic theory or supporting evidence for the Government's belief that by deflating demand they will bring inflation permanently under control and thereby induce an automatic recovery in output'. It also claimed that the monetarist policy would deepen the depression, erode the industrial base, and threaten social and political stability. Other policies (by implication, Keynesian) should be tried.

The Impact of Sir Keith Joseph

Also important in changing the climate has been the role of political activists who have 'political leverage' and are willing to push their ideas. One thinks of Joseph Chamberlain tying his own considerable political reputation to promoting the ideas of protective tariffs in the first decade of this century, of Enoch Powell almost single-handedly injecting immigration on to the agenda of British politics in 1968, or of Tony Benn taking up the cause of Labour party institutional reform after 1979. Such policies did not of course originate with these politicians. But their adoption of the ideas gained them more media attention and placed them on the agenda; the political weight of the politician and his supporters was attached to the issue. If he or she is a political leader then the party may adopt the idea. The emergence of Mrs Thatcher and Sir Keith Joseph to commanding positions in the Conservative party in 1975 and their support for many of the above ideas are an important part of the story of change in recent British politics. Enoch Powell was the major spokesman for the market economy in the early 1960s but once he turned to immigration he lost some of his audience. In his period of rethinking in 1974 Sir Keith was influenced by his reading of Friedman and

Hayek about the monetary causes of inflation and in turn persuaded Margaret Thatcher.

Sir Keith Joseph was the first Conservative front-bench figure to offer a sustained and broad-ranging challenge to the direction of post-war British economic management and to the record of the 1970–4 Conservative government. This was significant because he had been a prominent member of that government and presided over a big increase in expenditure at the Department of Health and Social Services. To critics of his change of mind he willingly admitted the error of his past ways; he had learned from his mistakes. His speech in Preston on 4 September 1974, on the eve of the second 1974 general election, came as a bombshell to the party and the public. It followed extensive Shadow Cabinet discussion of his ideas and his failure to persuade his colleagues there about the need to change course in fighting inflation. The speech contained a repudiation of the role of incomes policy (on which the Conservative party had fought the previous general election and which it still officially supported) and a criticism of his government's reliance on increasing demand to offset unemployment. The speech was a subtle call for more stringent control of the money supply and claimed that, if this was not done, there was no effective way of controlling inflation. Yet it was easy to read the speech as 'Joseph calls for more unemployment to reduce inflation.' In a series of subsequent speeches Sir Keith spelt out the case for a social market economy and monetarism. But monetarism was not enough: there was a need also to cut public spending and reduce state employment. In his Preston speech he said:

> The monetarist thesis has been caricatured as implying that if we get the flow of money spending right, everything will be right. That is not—repeat not—my belief. What I believe is that if we get the money supply wrong—too high or too low—nothing else will come right.

These speeches were given added significance when Mrs Thatcher, as leader, gave him responsibility for supervising the party's research and policy development.

In the speeches of Sir Keith during 1974–5 one finds the nearest statement to a 'New Right' credo.[17] Government is granted a role, primarily to create and uphold a framework of law and services which permit people to make the maximum number of decisions for themselves. He claimed that not all unemployment is involuntary and that the official unemployment statistics actually exaggerated the number of those genuinely seeking work. There were different categories of the unemployed: the workshy, the unemployable, those engaged in the black economy but registered as out of work, and those in transition between jobs were to be distinguished from those genuinely seeking employment. The trouble was, Sir Keith Joseph insisted, that so often government based its management of demand on the 'unadjusted' unemployment figures. Keynes was wrongly invoked in defence of these policies; in fact, Keynes was less concerned about expanding demand *per se* than about getting a better distribution between demand and supply. Expanding demand was no answer for frictional and other sorts of unemployment not due to demand deficiency. Government efforts to expand demand had led, Sir Keith claimed, to increased inflation and provided only temporary palliatives for the unemployed. Governments usually fell back on wage and price controls to control inflation and distorted the natural working of a market economy.

Sir Keith Joseph was also critical of the so-called 'middle way' of collectivism, which had resulted in extensive state regulation, high taxation, high levels of public spending, borrowing, and subsidies, and Keynesian demand-management. He urged a return to a policy of 'continence and smaller deficits'. The middle way had been pursued because politicians believed that most voters favoured it and that it was a way of providing for policy continuity and consensus. But the middle ground was an illusion, according to Sir Keith, and was not a mid-point between Labour and Conservative positions. Each time a Conservative government, in the interests of policy stability, accepted Labour policies which it had earlier opposed (for example, public ownership, high public spending, or comprehensive education), Labour moved to new left ground and so the middle-point moved with it—the so-called ratchet effect. Joseph urged his party to find the common ground, one which was supported by most of the electorate. He was sure that this was on the right of the middle ground and nearer to Conservative than Labour policies. The Conservative party had in the past compromised too much with socialism.

Sir Keith also claimed that an excess of socialism (usually equated with the pursuit of equality of outcome, nationalization, and state intervention) had overburdened the economy. This was a favourite New Right theme: the ways in which the productive wealth-creating private sector is drained by the public sector—'the lean kine'.

Sir Keith Joseph during these years was clearly hungry for ideas, eagerly studying economics textbooks and other works, and picking the brains of academics and independent scholars. He explained his new thinking to William Keegan: 'I understood from the Alan Walters and Peter Bauers of this world that deficit financing and borrowing are one of the main causes of our troubles.'[18] This was wedded to the enthusiasm of the convert or the born-again Conservative. In a remarkable statement, Sir Keith claimed that he only became a Conservative in 1974: 'But it was only in April 1974 that I was converted to Conservatism. I had thought that I was a Conservative but I now see that I was not really one at all.'[19]

How Mrs Thatcher challenged many of the operating ideas and practices of her Conservative predecessors is discussed further in Chapter 9. They had been accommodating and defensive, believing it politically necessary to maintain the welfare state, full employment, and high levels of public expenditure and to conciliate the trade unions. For much of the post-war period successive Prime Ministers had felt that the electoral advantage lay in leading their parties from the centre of the party. The art was to persuade their activists that they were being true to their principles while convincing the electorate that they were following policies which were non-doctrinaire and in the national interest. Mrs Thatcher's beliefs in individualism, thrift, and self-reliance are not very different from those of Mr Heath in 1970, but her sense of conviction is. Government, she believes, has a limited capacity to do good, but a great capacity to do harm, not least by distorting or interfering with the 'natural' working of society and the market economy. In contrast to Baldwin and the immediate post-war generation of Conservative leaders she lacks the sense of what she termed 'bourgeois guilt' (perhaps aristocratic is more appropriate) for the mass unemployment of the 1930s.

Between 1974 and 1979 a number of other opinion-formers were moving to the right. Most were successful journalists or had platforms which enabled them to endorse many New Right themes. Hugh Thomas, for example, had resigned from the Foreign Office in 1956 in protest at the British invasion of Suez and had joined the Labour Party. He became an academic and a best-selling historian, moved to the right and became an adviser on foreign policy to Mrs Thatcher. In due course he was ennobled by her and became Director of the CPS. A number of prominent former Labour supporters who deserted in the 1970s and joined the Conservative party expressed their views in *Right turn: Eight Men who Changed Their Minds*, edited by the Tory MP Patrick Cormack in 1978.[20]

Conclusion

The available survey evidence indicates that the electorate became more Conservative after 1974 even as the party moved to the right. It is true that there has long been clear support for Conservative positions on populist–authoritarian issues like law and order and immigration and some growing hostility to nationalization and strikes. David Robertson has recently monitored changes of opinion on many questions that were asked in identical form in October 1974 and May 1979. Of seventeen questions that can be directly compared the electorate moved to more right-wing average positions on fifteen and to the left only on the question of increasing cash to the NHS. He claims, 'the brute evidence is that the rightwards shift of the Conservative policy in the 1970s was at least matched by a similar shift in mass opinion'.[21] This was paralleled by a significant decline in support for many established Labour policies among party voters between 1964 and 1979. But surveys showed that the major programmes of the welfare state still attract strong support; most voters prefer to see their expansion even at the cost of rising taxes.

These findings have also to be set alongside the lack of popular appeal which capitalism has for the public and the apparent political weaknesses of private business. A survey by Marplan in April 1983 found that socialism was a more positive symbol than capitalism. Capitalism is popularly regarded as freedom of private business from government interference and socialism as equality and fair standards of living. Negative comments on capitalism (for example, inequalities of money and power) outnumber such views about socialism (too much state control or Marxism). Interestingly, more people spontaneously identify socialism with the Labour party (11 per cent) than capitalism with the Conservatives (1 per cent) and Labour politicians more frequently proclaim themselves socialists than Conservatives describe themselves as capitalists. Rose comments on the findings: 'The Conservatives have been far readier to describe themselves as anti-Socialist, than to say what they stand for.'[22]

Private business in Britain has not been a very effective lobby for a long time. For many years British economic management placed 'sound' finance—whether it was sustaining the value of sterling before 1931, correcting a balance of payments deficit or attacking inflation in the post-war period—ahead of economic growth.[23] Since 1979 the dramatic decline of manufacturing industry has proceeded with little effective protest from its spokesman, the CBI. The Conservative government has clearly preferred to have a high exchange rate, in large part to help the battle against inflation and to stiffen employer resistance to higher pay settlements, even if this has made British exports less competitive. As Colin Leys notes, the diversity of the CBI's members (between manufacturing and non-manufacturing interests, public and private sector groups, and firms largely reliant on export or on home markets) weakened its ability to speak with one voice. But its fear of the leftward shift in the Labour party also made it more tolerant of Conservative economic policy which did not promote manufacturing interests.[24]

In the United States there has been a similar, indeed more powerful, revival of libertarian free-market and moralistic ideas. Think-tanks and policy research bodies like the Washington based American Enterprise Institute, Hoover Institution for the study of War, Peace and Revolution (at Stanford,

California), and Heritage Foundation are well funded (the AEI alone spent $15 million compared to a total of some $2 million among the British groups covered in this chapter). Heritage was founded in 1973 and is backed by money from the Mellon family and Joseph Coors, the brewers. They are also intellectually respectable, politically influential bodies and have been able to draw on distinguished academics (such as Friedman and George Stigler) and people with government experience (such as Jeane Kirkpatrick at AEI). They aim to influence both the climate of opinion and particular policy questions. Heritage, for example, provides policy briefs for Washington politicians, consultants, and media reporters. In addition there are political organizations, such as the Conservative Caucus, the National Conservative Political Caucus, and the Committee for the Survival of a Free Congress. These have exploited the new technology of television and computerised mailing to build up lists of supporters and raise enormous funds. To some extent the groups have supported the many single-issue pressure groups which are sympathetic to right-wing causes, for example the anti-abortion group Life Amendment Political Action Committee (LAPAC) and Gun Owners of America.[25]

In the United States the opinion-forming and policy-implementing roles of these bodies have been helped by various factors, including the profusion of radio and television channels, opportunities to purchase air time on the channels and the sheer variety of newspapers and journals which help them gain a platform. Their ideas are discussed sympathetically in journals like the American *Spectator*, *Public Interest*, and *Commentary*. They are backed by powerful lobbies like the Moral Majority, National Taxpayers' Union, and various tax-cutting offshoots of the successful Proposition 13 campaign in California. They have also been associated with an electorally successful President in Ronald Reagan and with the economically dynamic and politically influential 'Sunbelt' of the West and South.

Such themes as hostility to major government intervention in the economy and welfare as well as to growing taxation, bureaucracy, and regulations on commerce and business, and support for a strong stand on defence and law and order are common to the right in both countries. But there are also important differences. Religious groups have been more readily mobilized in the United States about such issues as abortion, pornography, school prayers, and feminism. The courts and particular judges have been attacked for an allegedly liberal bias on many 'new politics' issues, like racial integration, school bussing, free speech, and rights of defendants in court. American groups have also been more concerned with a foreign policy that is not too 'soft' towards the Soviet Union, nor in Latin America, nor in the United Nations.

Bosanquet has claimed that the ideas of the New Right have had a greater impact in Britain and the United States than in Western European countries because of the relative weakness or absence of strong social democratic or christian democratic traditions in the first two countries. In the United States Conservatives have also been able to invoke a liberal consensus which includes a hostility to government.[26] In Britain the combination of relative economic decline, which has undermined many established political leaders and political and economic ideas, and the decline of the Labour party provided the New Right with its opportunity. In both countries the groups have satisfied Ashford's conditions of intellectual respectability, media support, approval of influential politicians, and electoral popularity.[27]

Notes

1. *After the New Right* (London, Heinemann, 1983).
2. *British Journal of Political Science* (1983), p. 193.
3. The most prominent spokesmen for libertarianism are Robert Nozick, *Anarchy, State and Utopia* (Oxford, Blackwell, 1975) and Murray Rothbard, *For a New Liberty* (New York, Collier, Macmillan, 2nd edn. 1972).
4. J. Burton, 'The Political Economy of the New Middle Way'; paper presented to Political Studies Association, Manchester, April 1985.
5. See P. Steinfels, *The Neo-Conservatives* (Chicago, Simon and Schuster, 1980) and G. Peele, *Revival and Reaction* (Oxford University Press, 1984).

6. P. Clarke, *Liberal and Social Democrats* (Cambridge University Press, 1978).
7. S. Brittan, *Left or Right: the Bogus Dilemma* (London, Secker and Warburg, 1968).
8. See S. Brittan, *Capitalism and the Permissive Society* (London, Macmillan, 1973) and *The Role and Limits of Government* (London, Temple Smith, 1983).
9. A. Seldon (ed.) (London, Institute of Economic Affairs, 1985).
10. (Harmondsworth, Penguin, 1980), p. 189.
11. *Conservative Essays* (Cambridge University Press, 1978), p. 150.
12. See B. Jessop *et al.*, 'Authoritarian Populism, two nations and Thatcherism', *New Left Review* (1984) and R. Levitas (ed.), *The Ideology of the New Right* (London, Polity Press, 1986).
13. 'The New Right, Social Order and Civil Liberties', in *The Ideology of the New Right*, p. 197.
14. *Capitalism, Socialism and Democracy*, p. 295.
15. *After the New Right*, Chap. 1.
16. 'The Political Economy of the New Middle Way'; paper presented at Political Studies Association conference. Manchester, 18 April 1985.
17. See his *Reversing the Trend* (London, Centre for Policy Studies, 1975) and *Stranded in the Middle Ground* (London, Centre for Policy Studies, 1976).
18. W. Keegan, *Mrs Thatcher's Economic Experiment* (London, Penguin, 1984), p. 46.
19. *Reversing the Trend*, p. 4.
20. (London, Leo Cooper.)
21. D. Robertson, 'Adversary Politics, Public Opinion and Electoral Cleavages', in D. Kavanagh and G. Peele (eds.), *Comparative Government and Politics* (London, Heinemann, 1984).
22. R. Rose, 'Two and a Half Cheers for the Market', *Public Opinion* (1983), p. 12.
23. W. Grant and D. Marsh, *The Confederation of British Industry* (London, Hodder and Stoughton, 1977).
24. C. Leys, 'Thatcherism and British Manufacturing: A Question of Hegemony'; paper read at Harvard University, April 1985.
25. G. Peele, *Revival and Reaction.*
26. L. Hartz, *The Liberal Tradition in America* (New York, Harcourt, Brace, and World, 1955).
27. N. Ashford, 'Conservative Ideas and Economic Policy'; paper presented at Political Studies Association, Southampton, April 1984.

5

BREAKDOWN

In Chapter 2 it was argued that many elements in the post-war consensus were part of a coherent package. No doubt there were potential contradictions between the different elements and some vulnerability in the overall package, and these became more visible during the 1970s. But for a quarter century or so after 1950 the policies were widely accepted by the élites and the dominant groups in the main political parties. This policy consensus coexisted with a much-praised political system, incorporating a dominant Prime Minister, effective government, a deferential populace, two parties evenly matched in the electorate, and very low levels of inflation and unemployment.

In this chapter we examine how the main values and policies identified in Chapter 2 lost their dominance from the mid-1970s. At times I return to an earlier period to identify the origins of the weaknesses which became more apparent later. There had long been minorities in both parties dissenting from the consensus. In face of a gathering economic recession, a slow-down of economic growth, and the clear repudiation of each party in government in the February 1974 and 1979 elections these critics found greater support.

In talking of relative economic decline in Britain one has to distinguish between two periods of concern. The long-term decline and loss of competitiveness by British manufacturing in the face of competition from Germany and the United States dates back to the turn of the century.[1] During the 1950s and 1960s, however, the concern was that Britain's economy was growing at a much slower rate than those of other western industrialized societies. In the 1970s economic growth in most western states slowed down because of the international economic recession, which followed the sharp increase in oil prices in 1973–4. Although all industrial countries had to spend more on oil and less on other goods Britain was once more falling behind. In the course of the 1970s unemployment doubled in OECD countries, but in Britain it trebled and British inflation rose twice as fast as the average in OECD countries.

Many of the ideas and policies identified with the post-war consensus gradually came to be regarded as causes and symptoms of Britain's decline. By the mid-1970s there was a large literature on such themes as 'Bankruptcy of Government', 'Second Great Crash', 'Crisis of Governability', 'Pluralistic Stagnation', 'Overloaded Government', and 'The Illusion of Governmental Authority', much of it concentrated on the British experience. This was a change from the old Whiggish assumption that the British had an unmatched capacity for government and from the almost universal admiration of the British political system! From being the exemplar of stable representative democracy, indeed the exporter of political institutions, Britain was widely regarded as the classic case of a democracy being badly governed.

A number of commentators claimed that democratic government, particularly in Britain, suffered from an excess of popular expectations and demands and lacked adequate authority and resources to meet these pressures. Government, in a word, was overloaded and the object of much popular dissatisfaction. Samuel Brittan argued that there were 'contradictions' in democracy, arising from the tendency of competing politicians to raise unfulfillable economic expectations among voters.[2] More generally, government was seen to suffer from organized social complexity; it was involved in too many dependency relations.[3] Achievement of its policies required the co-operation of many other factors and the more it depended on these the more easily could things go wrong. The difficulty for government was that the range of problems for which it accepted responsibility grew while its capacity to deal with them decreased. It was also possible to criticize these trends as a crisis of social democracy—a regime of active government, high public spending and taxation, and voters looking instrumentally towards the government for benefits.[4] Britain was becoming harder to govern. Some of this gloomy analysis was a questionable extrapolation of current trends, some of it pertained to western democracy in general, rather than just Britain, and survey evidence was at best ambiguous concerning many of the claims about popular attitudes.[5]

Full Employment

There was no doubt about the success of economic policy in delivering very high levels of employment. Compared to the inter-war years, when it rarely fell below 10 per cent, it rarely exceeded 2.5 per cent in the thirty years after 1945. Professor W. A. Phillips was able to establish a relationship between rates of inflation and unemployment for the 1950s showing that an increase in the pressure of demand was associated with a rapid rise in prices and employment.[6] One could therefore 'trade off' price stability against economic expansion; an increase in unemployment could be traded for a fall in inflation and vice versa. In other words, politicians could manage the economy and choose between differing amounts of inflation and unemployment. But in the late 1960s it was clear, at least in Britain, that the economy could have increasing amounts of both; the term 'stagflation' referred to the combination of low growth with rising unemployment and inflation.[7] Moreover, research showed that it was taking larger injections of demand to reduce unemployment at each stage of the economic cycle; at each stage the relationships between levels of unemployment and inflation were deteriorating (see Table 5.1). Reflation by increased government spending not only caused inflation to accelerate but eventually left unemployment higher at the next downturn in the cycle. Keynesian economics faced something of an intellectual crisis.

Defenders of Keynes have made various claims on his behalf. One tactic has been to rescue him from vote-seeking

TABLE 5.1 *The Worsening Economic Performance*

Government	Average annual increase in retail prices %	Average number unemployed, of UK adults, seasonally adjusted
Conservative 1951–64	3.5	330,000
Labour 1964–70	4.5	500,000
Conservative 1970–74	9.0	750,000
Labour 1974–79	15.0	1,250,000

politicians. His most recent biographer has argued that post-war British politicians have overloaded the economy, in the sense of expanding public spending too fast. In an era of full employment and economic growth, one might expect budget surpluses—to offset inflation—as often as budget deficits. In fact, most western states practised 'one-eyed Keynesianism', budgeting for deficits.[8] In Britain, for example, deficits were registered in eighteen of the twenty-five years between 1951 and 1975.

Keynes was an élitist; his early, privileged years at home, his educational background at Eton and Cambridge, and then his working life at the Treasury and Cambridge University, led him to take for granted the authority of government and the role of disinterested political leadership. Critics have fairly seized on his assumption (shared with Beveridge and the Webbs) of 'the agency of a benevolent state serviced by a technocratic élite'.[9] It is interesting that Keynes's reading of Hayek's *Road to Serfdom* alerted him to the dangers of collectivism and planning to personal liberty. In 1944, shortly before he died, he wrote to Hayek:

> Moderate planning will be safe if those carrying it out are rightly oriented in their own minds and hearts to the moral issue . . . dangerous acts can be done safely by a community which thinks and feels rightly, which would be the way to hell if they were executed by those who think and feel wrongly.[10]

A second tactic has been to say that his ideas have been misused. Both Sir Keith Joseph and Mrs Thatcher have tried to distinguish Keynes from the post-Keynesians, or economists and politicians who have added their own gloss to his ideas. There is an element of body-snatching here, but politicians were indeed selective in their interpretation of Keynes (surely the fate of most theories which are vulgarized when used in the real world by civil servants and politicians). Keynes was not indifferent to inflation, money supply, or problems which collectivism and active government might pose for individual liberty. Mrs Thatcher has regularly tried to reclaim Keynes as a spokesman for sound finance, as in her speech to the 1984 Conservative Party Conference and in frequent jousts in the House of Commons. Sir Keith Joseph claimed in his speeches between 1974 and 1977 that 'what was said and done in his (Keynes) name has been quite different', and 'from what Keynes wrote it seems likely that he would have disowned most of the allegedly Keynesian remedies made in his name and which have caused so much harm'.[11]

Governments had traditionally expanded demand when unemployment was rising. A side effect of injecting monetary demand at a rate above the rate of productivity growth was higher inflation which governments, in turn, tried to control by resorting to incomes policy and deflation. In 1972 and 1973 the Heath government, still committed to full employment, expanded money supply and reflated the economy when faced by rising unemployment.[12] This was accompanied by an incomes policy in 1972 and 1973 to control inflation. The great fear in the mid-1970s was inflation. In 1974 retail prices rose by 19 per cent, wage rates by 29 per cent, and industrial production fell by 3 per cent (all post-war records); by August 1975, the price rise over the previous eleven months was a new record—26.9 per cent in spite of, or because of, the Social Contract. In March 1975 Denis Healey's budget abandoned the commitment to plan for full employment, choosing to cut instead of increase the deficit, a historic breach with one of the main planks of the post-war consensus.[13] With unemployment at over one million, the budget would have been expected to reduce taxes and/or increase public spending. In a speech to his party conference in 1976 the Labour leader, Mr Callaghan, boldly proclaimed the new thinking in a passage widely thought to have been written by his son-in-law, Peter Jay, the British Ambassador to the United States:

> We used to think that you could just spend your way out of recession and increase employment only by cutting taxes and boosting government expenditure . . . it only worked by injecting bigger doses of inflation into the economy followed by a higher level of unemployment at the next step . . . The option [of spending yourself out of a recession] no longer exists.

This began something of an intellectual revolution in economic policy. By 1976, after the IMF rescue package of sterling, there was a new economic regime of pay policy, cutting taxes and public spending, lowering the public sector borrowing requirement (PSBR), and setting cash limits for much central and local government spending and targets for M3 money supply. Unemployment doubled in the lifetime of the 1974–9 Labour government to 1.2 million or 6.2 per cent. The idea of a government spending its way out of recession became less convincing in the 1970s. British industry did not react quickly enough to expansionist policies; extra demand was largely spent on goods imported from abroad (so causing a balance of payments problem) or simply chased the existing amount of goods available (so aggravating inflation). In fact it was the Keynesian view of how the economy *as a whole* worked that was under attack. Increased public spending—only one aspect of Keynesianism and only one way to boost demand—became the public whipping boy. This reaction created the opportunity for monetarists, who offered an anti-inflation package. Controlling the money supply was a task which the government could perform, providing that it was resolute enough—and there was sufficient confidence among economic agents in government's resolution—not to print more money even when unemployment was rising.

The growing importance of economic management increased the influence of economists on governments. But the old Keynesian dominance was no longer unquestioned. By the mid-1970s it was possible to distinguish various voices among the economics profession about the best mix of policies to achieve the objectives of high employment, stable prices, and a satisfactory balance of payments. There were the neo-Keynesians of the National Institute of Economic Research who wanted to curb inflation by an incomes policy, while still pursuing Keynesian policies of economic growth. A second group, commonly identified with the Department of Applied Economics at Cambridge, argued that Britain's manufacturing was uncompetitive and wanted to protect the home market by import controls and, if necessary, an incomes policy, in order to suppress inflationary tendencies when demand was boosted. The group's call for import controls achieved some influence in the left wing of the Labour party and was incorporated in its alternative socialist strategy. Finally there was a growing school of monetarists who argued that the key to reducing inflation was a reduction in the money supply. They claimed that there were structural causes of unemployment and that if it was treated as if it were caused by insufficient demand it would only aggravate inflation. If wage bargainers pushed for inflationary wage increases (for example, at a level above that of output) this would result in an increase in unit costs, a decline in competitiveness, a loss of markets, and the destruction of jobs. It was also assumed that people would then change their behaviour and bargain 'responsibly'. Yet Keynesianism was still dominant in the economics profession and monetarism attracted the support of only a few. As noted, in March 1981, 364 economists, including most of the distinguished academics and members of the major university departments, attacked monetarist policies in a letter to *The Times*.

Trade Unions

The position of the trade unions also changed. By 1970 the Labour government had rejected the two features of what Moran calls 'compulsory collectivism', that is, incomes policy and legal regulation of the unions.[14] The new Conservative government in 1970 opted for reform of the unions, largely as an alternative to the first, although by the end of its period of office it had reversed the emphasis. The failure of the Conservative government's Industrial Relations Act of 1971, coupled with the harm it did to industrial relations (the number of working days lost due to strikes reached 23 million in 1972, the highest figure since 1926) and the damage done to the government's relations with the unions, suggested the futility of such a measure. Reforming (or limiting) the legal privileges of trade unions and statutory incomes policies seemed to have been taken off the political agenda. Gaining the consent of the trade unions seemed to the be key to social and political stability in Britain. In 1974 Labour won two general elections, largely on the claim that it could get on with the unions.

Yet the power of the unions, as it was termed, was more accurately described by Peter Jenkins as the Social Democratic Dilemma.[15] He posed the question: how was the Labour movement to reconcile the political objectives of the trade unions, that is, maintain full employment, the welfare state, and stable prices with its industrial objective of free collective bargaining? In 1942 J. A. Schumpeter, while anticipating the eventual replacement of capitalism by socialism, had noted that 'the real problem' for socialism would be the position of organized labour. He wrote: 'A government that means to socialise to any great extent, will have to socialise the trade unions. And as things actually are, labour is of all things the most difficult to socialise.'[16]

British governments tried various approaches to curb inflation in the 1960s and 1970s—voluntary and statutory incomes policies, greater legal regulation of industrial relations, bargaining or a social contract approach and, finally in 1975, the abandonment of a commitment to full employment. Ministers oscillated between burden sharing, social and political partnership, persuasion, and coercion. From 1961 governments periodically tried to sustain Keynesian policies by incomes policies that led them to seek the co-operation of the union for wage restraint. Most of the various types of incomes policies lasted for two to three years before breaking down in the face of reaction against rigidities, compressed differentials, and growing disparities in incomes between public and private-sector workers (Table 2.3). It matters little whether the process is called corporatism, neo-corporatism, tripartism, or whatever; governments sought to bring groups into the process of policy-formation to limit market forces in settling the levels of pay and prices. John Goldthorpe describes the process as one in which 'partial substitutes are sought for market mechanisms, in order to aggregate and convert different interests and to render their pursuit consistent with policy goals'.[17]

The so-called Winter of Discontent was a rash of strikes, often unofficial and largely in the public sector, against the Callaghan government's attempt to impose a 5 per cent norm for wages and salaries. In December 1978 the Ford Motor Company eventually settled a strike at the cost of a 15 per cent wage rise and a similar figure settled a strike of BBC technicians. In the New Year there was an outburst of strikes and militant picketing by lorry drivers, ambulance drivers, oil-tanker drivers, and local government manual workers which resulted in the closure of schools, disruption of hospitals, and even in one well publicized case a refusal to bury the dead. The government and many union leaders seemed powerless to act. All this was graphically reported by the mass media. The unions reached new heights of unpopularity with the public and Labour's claim to have a 'special relationship' with the unions was destroyed.

One route to redistribution is via social policy and welfare expenditure—the Beveridge package. Another is via free collective bargaining. The Labour movement in Britain has been reluctant to acknowledge the tension between these two approaches. A good example is seen in the objections of some trade union leaders to the introduction of a scheme of child benefits in 1977 when they realized that the benefits would be paid directly to the mother and deducted from the wage packet. On the whole the results of free collective bargaining have not been kind to those who want to promote redistribution or greater equality of incomes. And the experience of the 1974–9 Labour government undermined the belief that the unions would or could trade off wage claims for an increased social wage. In various ways the Winter of Discontent in 1979 may have spelt the bankruptcy of a political tradition and style of government in Britain. Governments had been wary of the alternative anti-inflation policy (stricter control of money supply) which achieved wage restraint through weakening the unions. They feared that the consequent higher unemployment which would follow a monetary squeeze would be politically unacceptable. By 1979, however, the new Conservative government's strategy for beating inflation explicitly repudiated a policy for wages and rested on the control of the money supply.

The end of 'partnership' between the government and unions also meant that ministers would be less disposed to make policy concessions to union leaders. The other 'social partners' had not been unmoved by the events of 1974–6, what Colin Leys calls a 'political crisis of capital'.[18] The CBI gradually reacted against the Social Contract drawn up with the TUC, which granted concessions to the unions. Measures to extend job security and the closed shop, the imposition of price and dividend controls, the promise of planning agreements between the state and large firms, and the appointment of the Bullock Committee to make recommendations about industrial democracy amounted to a major incursion on managerial authority. CBI leaders became more overtly sympathetic to the Conservative party. They tried (and failed) to establish a strike insurance fund in 1978–9 and set up a 'Steering Committee on the Balance of Power'. (The ideas of the latter body were substantially reflected in the subsequent Employment Acts of 1980 and 1982.)

Welfare

The performance of the welfare state was attacked from both right and left flanks. There were differences among Conservative critics but all professed concern about the mounting costs of the various programmes, the drain on productive investment, and the failure to concentrate resources on the most needy. They therefore favoured more selectivity in the distribution of welfare resources. Other Conservatives, echoing the traditional concerns that had always opposed *L'État providence*, worried that the welfare state weakened the work ethic and lessened the recipient's sense of personal reliance and responsibility. Indeed in the 1970s there developed the so-called 'why work?' syndrome; for some low-paid workers, whose welfare benefits could equal or exceed what they might earn, or for whom an increase in earnings would be wiped out by loss of benefits and tougher taxes, it made financial sense to stay unemployed. Other critics simply objected to the state playing such a large role and favoured the private provision of services. The free-market spokesmen, led by the Institute of Economic Affairs, urged that users of state services should be charged in full, that a reverse income tax should be introduced to help the poor, and that the state should confine itself to providing public goods. The critics were, variously, cost-cutters, selectivists, moralists, and privatizers.

Conservative supporters of the welfare state pointed out that it promoted co-operation and social harmony and refuted the idea that the welfare state was parasitic on the industrial system by referring to successful economies in Sweden, Norway, and West Germany all of which spent higher shares of their national incomes on welfare. Conservative leaders feared the likely electoral unpopularity of adopting measures to cut state provision of welfare for tax cuts. One senior official dismissed such claims with the remark, 'You can't shoot Santa Claus.' Conservative ministers regularly boasted of their spending achievements on welfare. Sir Keith Joseph, then Secretary of State for Social Services, told his party conference in 1973 that real spending on the NHS and social services was rising at an annual rate of 40 per cent more than under the 1966–70 Labour government: 'In other words, we are spending more in real terms than Labour ever managed to do.'

From the other flank, Titmuss and his colleagues (Townsend and Abel Smith) claimed that poverty still existed on a large scale and was proof of the failure of welfare policies. They defined poverty in relative terms, raising the poverty line when living standards increased. By linking the poverty line to a level of 40 per cent over national assistance levels of benefit, Abel Smith and Townsend claimed that in 1960 14 per cent, or 7.5 million, actually lived in poverty.[19] As more people claimed means-tested benefits, so the numbers relying on supplementary benefits increased. Later Townsend, in his *Poverty in the United Kingdom*, claimed that a quarter of households lived in poverty.[20] During the 1970 general election campaign the Child Poverty Action Group claimed that the conditions of the poor had actually deteriorated under the Labour government. The Fabian Society also published highly critical reports of the 1964–70 and 1974–9 Labour governments' records on equality.[21] The failure of the welfare state to eliminate poverty led Townsend to doubt the ability of parliamentary socialism to produce 'radical structural change' in society and to claim that 'democratic socialism did not fail in the 1960s, it was not tried'.[22] He concluded that 'the fundamental question left unanswered by Labour's rule is whether democratic socialism can be effective'.[23]

Marxist and neo-Marxist writers claimed that the slowdown of economic growth was producing a new 'contradiction' in capitalism. Advanced capitalist societies required a high level of welfare spending to gain popular support or compliance but, increasingly, this was achieved at the cost of squeezing the profits of industry. Hence the 'contradiction' or fiscal crisis of the capitalist state arising from the conflicting needs of maintaining political support and providing for capital accumulation.[24] This view, interestingly, joined with right-wing claims that welfare was being placed ahead of economic growth and that high taxes and egalitarianism were inimical to creating a prosperous economy.

Critics also pointed to the failure of existing policies to produce a more radical redistribution of life opportunities. Much data showed that many programmes of public expenditure did not promote equality: the middle class were able to do well because of their skills and ability to work the system, and gained from the pattern of state expenditure and tax concessions in housing, higher education, and rail communications. For example, the top fifth of income earners received nearly three times as much public expenditure in education per household as the poorest fifth. One verdict was that 'in some cases, it is likely that there would be greater equality if there was not public expenditure on the service concerned'.[25]

There was some concern about the lack of visible political support for the welfare state. Welfare services are labour intensive and rarely have a return which can be regarded as productive. At the same time the egalitarian welfare assumptions, enshrined in the Butskellite consensus, were under strain because of slow growth. We do not have good evidence about popular attitudes to the welfare state but a European Community study conducted in member countries in 1976 showed the British to be rather tough-minded in comparison to people in other countries. For example 36 per cent of the British sample thought the government was not doing enough for the poor, compared to an average of 54 per cent in other EEC states, and 36 per cent thought that people in Britain were living in real poverty compared to an average of 47 per cent in other EEC states. British respondents were more likely to attribute poverty in their country to personal failings of the poor rather than to structural reasons like unemployment, which other Europeans mentioned. It may be that low economic growth in the 1970s, far from promoting solidarity or radicalizing people, actually dented altruism—a crucial motive for the values of welfare and redistribution. Altruism was found more frequently among the affluent, the middle class, and the better educated.[26]

The character of trade unionism in Britain may also have militated against redistribution. Britain's working class may be the most 'mature' in Europe in the sense that, because Britain was the first industrialized state in the world, it is the oldest and has become increasingly self-recruiting. But the working class is also highly differentiated and this has increased any government's (particularly a Labour government's) problems of offering policies which do not divide it. It has become increasingly divided into the affluent and non-affluent, public and private-sector employees, skilled and unskilled workers, home owners and council tenants, and members of strong and weak trade unions.

The advance of the welfare state has depended greatly upon economic growth, perhaps more than upon political ideology.[27] Yet socialists as diverse as Karl Marx, Ramsay MacDonald, Anthony Crosland, and Harold Wilson (at least before 1964) all assumed that the problem of production was about to be solved. In *The Future of Socialism*, written in 1956, Anthony Crosland envisaged a future period when this might indeed be the case. Then, he wrote, he might be prepared to 'stop worrying about hard work and economic matters and to relax into greater leisure and more cultural pursuits'.[28] As a Labour Cabinet minister he later recognized the indispensability of economic growth for the realization of his goals. A Labour government, like any other, was dependent on a thriving mixed economy to provide a surplus for welfare. Yet its commitment to socialism placed it in the position of seeming to fatten the golden goose only to kill it off eventually. In the short term the necessity to pursue policies which would encourage business and finance confidence depressed its more ardent followers. Its dilemma was whether to make capitalism work and deny its 'socialist' ideology or work to undermine capitalism and face turmoil in the short term. The latter might be functional to the achievement of socialism but it made it difficult to win elections.

In the 1970s the onset of economic recession with high unemployment and inflation meant that complaints about the rising costs of welfare were heard more frequently throughout Western European states. Conservative leaders regularly attacked the 'false gods' of egalitarianism, envy, and the swollen and non-productive public sector. Heclo notes that western governments did not try to dismantle the welfare state in response to these pressures. Instead they slowed down the rise in public spending, were reluctant to introduce new programmes, sought greater value for money, and trimmed programmes.[29] Surveys suggested that there was a willingness to reduce expectations and that levels of life satisfaction remained high in spite of economic disappointments.[30] Contrary to the fears of some Conservatives the welfare state had not weakened liberty. It is more correct to say that the growth of welfare after 1945 coincided with an expansion of personal liberty and opportunity for many. In view of the contribution of welfare to the stability of regimes, it was perhaps strange that some right-wing Conservatives should attack such an admirable conservative institution.[31]

Active Government

The problem of 'big government' loomed large in the rhetoric of right-wing politicians in the 1970s and the new cry was for rolling it back. The growth of government can be measured along many indicators—taxation, public spending, programmes, state ownership of industry, state employment, and so on, and we should carefully distinguish between these different indicators. Some part of the right-wing attack on 'big government', bureaucracy, and high levels of taxation (a regular theme of successive Conservative election manifestos) was not merely rhetoric but also a reaction to the economic recession and the slow-down of economic growth in the 1970s which posed problems for funding government programmes. In a slow-growth economy like Britain's, pressures for increases in public spending have collided with pressures from workers to protect take-home pay. In the thirty years' period 1951 to 1980 the British economy grew by an average of 2.5 per cent annually, public spending by an average of 4.6 per cent, but take-home pay by only 1.7 per cent.[32] There is some evidence that the 'wage capping' effect was resented by well-organized groups of workers and led to industrial disruption.[33] Comparative research also suggested that the political protest or tax backlash movements were strongest in countries where direct or more visible taxes increased the most.[34]

There was a predictable reaction to the failures of government. Commentators and politicians urged people to lower their expectations, particularly over economic benefits. Sir Keith Joseph argued that the commitment to maintain full employment should be abandoned on the grounds that a government in a free society could only create the conditions for enterprise to flourish; it was up to employers and workers to deliver goods and services which customers were willing to purchase. The public mood did change. If, during the 1950s and 1960s there was a strong correlation between changes in unemployment levels and the government's popularity, this has virtually disappeared since. Steep increases in unemployment since 1974 have not been fatal for a government's popularity. Hence politicians had 'a vested interest in promoting realistic, even cautious plans for the future growth of government programmes'.[35]

In line with this thinking was the tendency for people to blame other factors than the government—for example, trade unions, world recession, rising oil prices, and personal failings—as a cause of high inflation and high unemployment in the 1970s. According to James Alt's analysis of survey evidence:

> In large measure, then, the story of the mid-1970s is the story of a politics of declining expectations. People attached a great deal of importance to economic problems, people saw clearly the developments that were taking place, and people expected developments in advance and thus were able to discount the impact of the worst of them. However, in unprecedented numbers, people also ceased to expect the election of their party to make them better off, largely because they also ceased to expect it to be able to do very much about what they identified as the principal economic problems of the time. The result of this—as well as perhaps some of the other factors discussed above—was not a politics of protest, but a politics of quiet disillusion, a politics in which lack of involvement or indifference to organised party politics was the most important feature.[36]

Mixed Economy

The parties in government continued to make marginal adjustments to the private–public sector boundaries. The Heath government sold off Thomas Cook's travel agency and the Carlisle state brewery was sold to the private sector. But (following its bankruptcy) Rolls–Royce was taken over in 1971 and the following year the government intervened to prevent the closure of the Upper Clyde shipyard. The 1974 Labour government's policies promised a major extension of public ownership and government intervention in the economy. Its Industry Act of 1975, though much watered down from the plans hatched in opposition, established a National Enterprise Board with a remit to extend public ownership and make voluntary planning agreements with firms. By 1979, however, the NEB had made only one planning agreement. The Labour government also rescued the lame duck British Leyland, sections of the aerospace and shipbuilding industries were taken into public ownership, and in 1977 the British National Oil Corporation was established. The party's *Programme* in 1976 proposed further nationalization measures and compulsory planning agreements. The old agreement, albeit reluctant, about the status quo regarding the public–private divide, was waning. Among the new Conservative leadership, Mrs Thatcher and her Treasury team were worried about public spending, assertive public-sector trade unions, and loss-making nationalized industries and promised a programme of privatization. At the same time Labour's left wing was as convinced as ever of the failings of British capitalism and more interested in establishing a powerful state holding company which would take over profitable and strategically important firms.

Social Engineering

There also appeared to be greater self-doubt among decision makers about their problem-solving abilities. Richard Rose has observed that at the beginning of the century governments were mainly concerned with the classic defining activities of the modern state, namely diplomacy, defence, law, finance, and public order.[37] Governments had reasonably well-established technologies for carrying out these functions. By contrast, not only are many activities of contemporary government more complex (be they the promotion of more social and economic equality, better health, economic growth, or educational standards) but governments also lack a social engineering capability for many of these tasks. Moreover, much data about social conditions and policies is ambiguous or indeterminate. When policies appear to fail—be it public ownership or monetarism—there are as many defenders calling for more of the same policies as there are opponents calling for their abandonment. And even where the information and policy knowledge is available the means may be politically unacceptable.

There has also been a new modesty about the possibility of social engineering. This is something separate from the traditional Conservative respect for the status quo and the Burkeian view of the complexity and independence of society, the wisdom inherent in existing institutions and practices, and scepticism about its amenability to manipulation by reason or naked intellect. Rather, the new modesty claims to have been born out of experience and the failures of successive wages policies, economic planning, regional policy, policies for economic growth, and high-rise council estates. All of this connects with Anderson's thesis of the Ignorance of Social Intervention in the sense that the intervener is unable to know what he is doing and predict that his actions will have intended effects.[38]

In the United States there was a similar disillusionment (though there have been disputes about their effects) with the various Great Society Programs to promote educational standards and equality and to combat poverty. Part of the negative reaction to these policies is to 'inputism', or the belief that complex social problems may be amenable to the investment of yet more resources. In the United States the authoritative Coleman Report[39] and the work of Jencks[40] demonstrated that differences in the input of resources and facilities had little or no, sometimes even an adverse, relationship to student achievements. Other factors, such as the commitment of teachers, the quality of the head teacher, or the cultural capital of the family seemed to be more important. Education appeared to be limited as a tool of redistribution, according to Jencks, because of deeply entrenched broader social and economic inequalities and the importance of the family.

One can readily see how much of this research fuelled fiscal conservatism in many countries. Interestingly, it has been a case of the monitoring role of social science being used to refute the problem-solving claims of the various professions. Critics were quick to point to the increased public expenditure and employment in services like health and education alongside little evidence of improved performance and claimed that voters had come to expect too much from politicians, who in turn believed that if the 'right' policies were adopted, then a wide range of economic, social, and racial problems could be solved. John Vaizey, in his study of a number of influential makers of the post-war consensus, commented that they were 'the first generation of Englishmen to believe that the solution to social and even political dilemmas lay in political action. This was the belated achievement of the Fabians, but, when it actually arrived, it was too late.'[41] As governments expanded their responsibilities, regardless of their ability to carry them out, so failure to satisfy the demands risked undermining confidence in the system. The growth of public expenditure (steadily commanding a larger share of GNP), the fiscal illusion that benefits could be provided for voters without regard to costs, and the increase in the number of pressure groups and lobbies dependent on the state for services and jobs all combined to expand the role of government. It was easy to conclude from this that the government's role should be reduced—a moral more welcome to Conservatives. Its 1979 manifesto said 'Attempting to do too much, politicians have failed to do those things which should be done. This has damaged the country and the authority of government. We must concentrate on what should be priorities for any government.'

Much concern centred on the growth of public expenditure. Under the PESC (Public Expenditure Survey Committee) system the expenditure figures for departments were adjusted to allow for inflation, so that money values were expressed in constant prices. But the upsurge of inflation in the 1970s played havoc with the system. The most important factor leading to the explosion of public spending was the assumption that spending commitments could be financed out of predicted rates of economic growth, which subsequently were rarely achieved. Much of the growth did not represent an increase in state power but took the form of transfer payments, such as pensions and various welfare benefits, which simply passed through the state's hands.

Public spending in Britain went out of control between 1973 and 1976, in the sense that final outturns far exceeded those predicted. Between 1973 and 1975 it increased from 51 per cent to 59 per cent of GDP, at factor cost (as public spending was then measured). In 1977 the government decided that only the part of the nationalized industries' capital expenditure which was covered by their internal resources would count as public expenditure. Public spending covered goods and services, such as the payments for labour in the public services and the costs of the services to the public and transfer payments in the form of payments and subsidies which the government makes to various people. In the post-war period the balance of public spending between different programmes had changed. Table 2.3. shows that between 1953 and 1973 spending on the popular social welfare programmes (of education, social security, health, and personal social services) grew from 32 per cent to 44 per cent. But on so-called defining activities (defence, law and order, foreign affairs, and repayment of debt interest) the figure fell from 45 to 27 per cent.[42]

In April 1976 a system of cash limits was established for about three-quarters of central government expenditure. Departmental allocations now included a figure for future levels of inflation and there was no guarantee that they would be increased if inflation was higher than calculated.

A more critical look at public spending was encouraged by the work of Bacon and Eltis, *Britain's Economic Problems: Too Few Producers*,[43] which described a shift in resources and employment from marketed to non-marketed goods and services. By 'market' they meant goods and services which are sold for a price and by 'non-market' they referred to goods and services provided by the government out of general revenue (such as state education, defence, and the National Health

Service). They argued that this shift reduced investment, employment, and growth in the market sector. The distinction is not quite one between 'unproductive' public and 'productive' private sectors, but it is near enough. Employment and spending had increased in the social services, health, education, and administration, while it had actually fallen slightly in the productive market sector. It was easy to conclude from this analysis that the growth of the public sector was:

(1) 'crowding out' investment and employment from the market sector;
(2) 'parasitical' on the market sector.

The thesis has been criticized for its inadequate appreciation of the productive side of public services (for example, the creation of a healthy and educated workforce or the role of the welfare state in improving the morale of the workers). Many public-sector goods are marketed (for instance, local government gains more than a quarter of its income from the sale of services). It also overlooked the fact that many of the new non-market workers were part-time and female, for whom the alternative was probably no work at all. The actual growth in employment in the public sector was rather modest, from 23.5 per cent to 26 per cent of the total workforce between 1964 and 1974, and was not out of line with what was happening in other western states.[44] It also reflected demographic pressures in the growth in numbers of old age pensioners and school-age children. The thesis could hardly be a sufficient explanation of Britain's decline, in view of the fact that similar trends were observed in other western states. But the analysis was influential. Two critics note: 'However, the work published by Bacon and Eltis in the late 1970s was exactly right for the political climate of the time.'[45]

The Conservatives found the analysis attractive for ideological reasons, given their negative view of the public sector and public spending. In her famous 'Let the Children Grow Tall' speech in September 1975 Mrs Thatcher drew attention to the steady growth in public-sector employment, while the overall working population contracted: 'So one must not overload it [the private sector]. Every man switched away from industry and into Government will reduce the productive sector and increase the burden on it at the same time.' In the Labour party also, influential ministers were coming round to this view. They were aware of resistance to higher taxes to pay for the social wage and wanted to contain the rise in social spending and leave more in take-home pay. The Public Expenditure White Paper in 1976 (Cmnd. 6393) complained about the growing public-sector borrowing requirement. It noted that in the three years 1974–6 public spending grew by 20 per cent in volume, and as a proportion of GDP from 50 to 60 per cent, but output had risen by less than 2 per cent. It continued: 'As recovery proceeds we must progressively reduce the deficit . . . more resources . . . will be needed for exports and investment.'

Political and Social Change

It is not surprising that, in view of the above developments, the two-party system came under growing strain.[46] The weakening loyalty of the electorate to the parties can be demonstrated in the increase in volatility as registered in opinion polls, by-elections, and general elections. In the twenty-year period 1945–65, there were only two years in which support for the two main parties varied by 15 per cent or more in the opinion polls. In the next twenty years there were only two years in which it failed to vary by less than that amount. There was also a marked decline in support for the two main parties at general elections. The Conservatives in the two 1974 elections and Labour in 1979 each descended to post-war lows in shares of the vote: for the latter it was the third successive election in which it failed to gain 40 per cent of the vote. Governments and party leaders plummeted to new depths of unpopularity, according to the opinion polls. The Labour and Conservative parties, which had normally gathered 90 per cent of the total vote at general elections between 1950 and the 1970s, fell to a combined share of 75 per cent in 1975 and 80 per cent in 1979. There was also a decline in the voters' identification with the political parties. From 1964 to 1979 the Conservative share of identifiers remained stable at 38 per cent, while Labour's declined from 43 to 38 per cent. But among Conservative identifiers those who claimed to have a very strong identification with the party fell from 48 to 23 per cent and among Labour identifiers the proportion of strong identifiers fell from 45 to 27 per cent (Table 5.2).

Disappointment at each government's economic record was one cause of the turning away from the main political parties. Governments presided over unprecedented levels of unemployment and inflation. The former doubled under the Labour governments from 1964–70 and 1974–9; the latter doubled under the Conservative government in 1970–4. There was a rise in support for other political parties, although the electoral system prevented this from being accurately reflected in seats in the House of Commons. The Liberals in particular suffered from the first past the post electoral system.

There was a dealignment of the two-party system with no clear signs of what a successor system might look like. The

TABLE 5.2 *The Incidence and Strength of Conservative and Labour Party Identification, 1964–1983*

	1964	1966	1970	1974*	1979	1983
Percentage identifying with:						
Conservative party	38	35	40	35	38	38
Labour party	43	46	42	40	38	32
Conservative identifiers only:						
Very strong identifiers	48	49	50	29	23	34
Fairly strong identifiers	41	39	40	50	53	43
Not very strong identifiers	11	12	10	21	24	23
	100	100	100	100	100	100
Labour identifiers only:						
Very strong identifiers	45	50	47	37	27	32
Fairly strong identifiers	43	41	40	44	50	39
Not very strong identifiers	12	9	13	19	23	29
	100	100	100	100	100	100

*The average of the February 1974 and October 1974 figures.

Note: The figures include respondents who at first said they had no party identification but who, in answer to a supplementary question, said they were 'a little closer' to one or other party.

Sources: 1964, 1966, and 1970 — election surveys conducted by David Butler and Donald Stokes for *Political Change in Britain*, 2nd edn. (London, Macmillan, 1974); 1974 and 1979 — British Election Studies, University of Essex; 1983 — BBC TV/Gallup survey, 8–9 June 1983.

decline of partisanship affected the whole electorate, but it proceeded faster among the well educated.[47] The class basis of the two-party system was also weakening. The 'normal' Conservative share of the middle-class vote, at four-fifths between 1950 and 1970, fell to three-fifths in the 1970s and Labour's 'normal' three-fifths share of the working-class vote fell to barely one half. Over time, affluence and social change was weakening the homogeneity of the working class. Many skilled workers, who were now home owners, were turning to the Conservative party. At the same time, more middle-class professionals, employed in the public sector and enrolled in trade unions, were turning to the Labour party. The blurring of class lines in voting spelt more long-term trouble for the Labour party than for the Conservative party. The decline in partisanship and the class basis of the party system, and Labour's unpopular stands on many issues left it more vulnerable. In spite of victories in the 1974 elections, Crewe, Svarlvik, and Robertson emphasized the erosion of support for Labour's policies among its 'core' working-class and union voters, and referred to the likelihood of 'long-term electoral decline if the majority of its supporters reject the majority of its basic policies. The position of the Labour party is likely to be particularly precarious after a spell in government in which it disappoints the economic expectations of its voters.'[48]

Social class was changing also. *Embourgeoisement*, or the assumption of middle-class lifestyles by manual workers, was a complex process. Yet, it seems clear that the spread of affluence and improved living standards contributed to a sharp fall in the proportion of voters perceiving important differences between the parties and in regarding the parties as representing the opposing class differences.[49] By 1979, 50 per cent of manual workers owned cars and 40 per cent of skilled workers were buying their own homes. This last proportion equalled the number of those who were renting from local authorities. The middle class expanded largely because the increase in service occupations attracted the sons and daughters of manual workers. By 1972 about a half of middle-class men came from working-class parents.[50] Yet it was possible to exaggerate the withering away of class.[51] The top 1 per cent of the population owned one-fifth of the country's wealth and the top 1 per cent of income earners received, after tax, the same as the total of the poorest one-fifth. These figures were little different from those of 1950. There were still large differences between the social classes in life chances.

Conclusion

This chapter has examined the dismantling of the main planks of the post-war consensus. It is interesting to refer back to Anthony Downs's analysis of why the two parties converged in policies in the 1950s and 1960s. How does one explain the Conservative party's subsequent move to the right and Labour's move to the left? In the case of Labour, left-wing activists were able to profit from disillusion with the Labour governments and the party's intra-party democracy and shift it to the left. Among Conservatives there was a reaction against the collectivist orientation of the Heath government as well as Labour's policies between 1974 and 1976. In Labour there was a call for a 'true' socialist strategy—not of responding to public opinion, as interpreted by opinion polls and the right-wing popular press, but of giving a socialist lead and having a radical programme for taming capitalism. In the Conservative party the resurgence of a 'politics of conviction' was allied to more right-wing policies. On issues like law and order, immigration, and trade unions the Conservatives could win votes by an authoritarian stance.[52] Sir Keith Joseph's point about a common ground (where most votes clustered) being different from the middle ground (a mid-point between the parties) was confirmed.

Questions of party structure are also important. In the case of the Labour party, the Party Conference, in which the political left had a more influential voice, came into its own when the party was in opposition. In the Conservative party, power is concentrated in the hands of a leader who has great authority to propound party policy. The character of Mrs Thatcher was decisive after 1975 and this is discussed in chapter 9. In both parties defenders of centrist policies lost ground and in both there were calls for a return to 'first principles'. Above all, the growing sense of economic decline placed those associated with the policies of the previous twenty-five years on the defensive. This unfolded against the worrying background of apparent government ineffectiveness—the miner's strike in 1973–4, inflation which increased 20 per cent per annum between 1974 and 1977, the resort to the IMF in 1976, and actual decline in the average citizen's standard of living in 1976 and 1977.

Economic weakness had political repercussions because the economy dominated the political agenda. The failure of politicians to carry out their promises fed back into the political system, as disappointed voters gradually withdrew their approval from one party then the other. This was often called the politics of economic decline and it spilt over to reduce the effectiveness of government.

It is interesting to consider two statements, among many, of what went wrong. One regards it as the failure of capitalism, the other as the failure of social democracy.

The 'crisis of capitalism' school is largely Marxist, or at least its views are propounded in left-wing circles. Ever hopeful of cataclysm, and counterposed to the revisionists in the Labour party, it claims that capitalism is inevitably prone to breakdown. It also argues that profits are squeezed to pay for wages and welfare, that is to purchase legitimacy, but at the cost of creating an investment crisis. Fulfilling the consent function of the modern capitalist state creates a fiscal crisis and gets in the way of its effectiveness.[53]

The 'crisis of social democracy' school, from both the political left and right, noted that Labour's right wing, or social democrats, had claimed that modern managed capitalism had changed and could deliver economic growth. Economic growth also lessened demands for state ownership, while funding public services and making the promotion of redistribution and equality easier. But the left could ask: what would happen if there was no economic growth? On past form there would be policies of deflation, 'sound' finance, wage restraint, and probably conflict with the trade unions. From a different political viewpoint Samuel Brittan argued that excessive popular expectations were produced by competing political parties and pressure groups and might result in political breakdown, because of 'excessive burdens being placed on the "sharing out" function of government'.[54] Other free-market critics also argued that the interventions of the government in the economy risked transforming economic and industrial disputes into potential constitutional crises.

Of these two analyses the second appears to have been the more persuasive to date. James Alt found in the 1970s that low economic growth, far from promoting solidarity or radicalizing people, has actually dented altruism—a crucial motive for the values of socialism and redistribution. Programmes of government intervention and public spending depend to some degree upon public confidence in the beneficence of government. As we shall see in Chapter 6 Labour found in 1979 and 1983 that this was not widespread.

Notes

1. S. Brittan, 'How British is the British Disease?', *Journal of Law and Economics* (1978).
2. 'The Economic Contradictions of Democracy', *British Journal of Political Science* (1975).
3. A. King, 'Political Overload', *Political Studies* (1975).
4. A. Wright, 'What Sort of Crisis?', *Political Quarterly* (1977).
5. See J. Alt, *The Politics of Economic Decline* (Cambridge University Press, 1979) and R. Rose and G. Peters, *Can Government Go Bankrupt?* (London, Macmillan, 1979).
6. A. Phillips, 'The Relationship between Unemployment and the Rate of Change of Money Wages in the UK, 1861–1957', *Economica* (1958).
7. M. Friedman, *The Counter-Revolution in Monetary Theory* (London, IEA, 1970).
8. Rose and Peters, *Can Government Go Bankrupt?*, p. 133.
9. R. Skidelsky (ed.), *The End of the Keynesian Era* (London, Macmillan, 1977), p. 62.
10. R. Harrod, *The Life of John Maynard Keynes* (London, Macmillan, 1957), pp. 192–3.
11. Sir Keith Joseph, *Reversing the Trend* (London, Barry Rose, 1975), p. 25.
12. M. Stewart, *The Jekyll and Hyde Years* (London, Dent, 1977), p. 138.
13. *Labour Party Annual Conference Report* (London, 1976), p. 188.
14. *The Politics of Industrial Relations* (London, Macmillan, 1977).
15. *New Statesman*, 20 September 1974.

16. *Socialism, Capitalism and Democracy* (London, Allen & Unwin, 1943), p. 379.
17. J. Goldthorpe (ed.), *Order and Conflict in Contemporary Capitalism* (Oxford, Clarendon Press, 1984), p. 12.
18. C. Leys, 'Thatcherism and British Manufacturing: A Question of Hegemony', paper read at Harvard University, April 1985.
19. B. Abel Smith and P. Townsend, *The Poor and the Poorest* (London, Bell, 1975), p. 71.
20. (London, Penguin, 1979.)
21. N. Bosanquet and P. Townsend, *Labour and Equality* (London, Heinemann, 1980), pp. 10–11.
22. Ibid., pp. 10–11.
23. Ibid., p. 229.
24. J. O'Connor, *The Fiscal Crisis of the Capitalist State* (New York, St Martin's Press, 1973).
25. J. Le Grand, *The Strategy of Equality: Redistribution and the Social Services* (London, Allen & Unwin, 1982), p. 32.
26. J. Alt, *The Politics of Economic Decline*.
27. P. Flora and A. Heidenheimer (eds.), *The Development of Welfare States in Europe and America* (New Brunswick, Transactions, 1984).
28. (London, Cape, 1956), p. 286.
29. H. Heclo, 'Towards a New Welfare State', in Flora and Heidenheimer (eds.), *The Development of Welfare States*, pp. 402–3.
30. R. Rose, 'Ordinary People in Extraordinary Circumstances', in R. Rose (ed.), *The Challenge to Government* (London, Sage, 1980).
31. A. King, 'The Political Consequences of the Welfare State', in S. Spiro & E. Yuchtman-Yaar (eds.), *Evaluating the Welfare State: Social and Political Perspectives* (New York: Academic Press, 1983).
32. R. Rose, *Understanding Big Government* (Cambridge University Press, 1985), p. 220.
33. F. Wilkinson and J. Turner, 'The Wage–Tax Spiral and Labour Militancy', in D. Jackson (ed.), *Do Trade Unions Cause Inflation?* (Cambridge University Press, 1975).
34. H. Wilensky, *The New Corporatism, Centralisation and the Welfare State* (London, Sage Political Sociology Series, 1976), vol. ii.
35. Rose and Peters, *Can Government Go Bankrupt?*, p. 249.
36. *The Politics of Economic Decline*, p. 270.
37. 'On the Priorities of Government: A Developmental Analysis of Public Policies', *European Journal of Political Research* (1976).
38. D. Anderson, *The Ignorance of Social Intervention* (London, Croom Helm, 1980).
39. J. Coleman, *Equality and Educational Opportunity* (Washington DC, Government Printing Office, 1966).

40. C. Jencks, *Inequality: a Reassessment of the Effects of Family and Schooling in America* (London, Allen Lane, 1973).
41. *Enemies of Promise* (London, Weidenfeld and Nicolson, 1983), p. 5.
42. R. Rose, *Politics in England* (London, Faber, 4th edn. 1985), pp. 364–5.
43. R. Bacon and W. Eltis (London, Macmillan, 1976).
44. M. Stewart, *The Jekyll and Hyde Years*, p. 225.
45. K. Newton and T. Karran, *The Politics of Local Expenditure* (London, Macmillan, 1985), p. 35.
46. S. Finer (ed.), *Adversary Politics and Electoral Reform* (London, Wigram, 1975) and R. Rose, *Do Parties Make a Difference?* (London, Macmillan, 1984).
47. I. Crewe, B. Sarlvik, and J. Alt, 'Partisan Dealignment in Britain 1964–1974', *British Journal of Political Science* (1977).
48. 'Partisan Dealignment', p. 187.
49. D. Butler and D. Stokes, *Political Change in Britain* (London, Macmillan, 1969).
50. J. Goldthorpe, *Social Mobility and Class Structure in Britain* (Oxford, Clarendon Press, 1980).
51. J. Westergaard and M. Resler, *Class in a Capitalist Society* (London, Penguin, 1975).
52. I. Crewe and B. Sarlvik, 'Popular Attitudes and the Conservative Party', in Z. Layton-Henry (ed.), *Conservative Party Politics* (London, Macmillan, 1979).
53. J. O'Connor, *The Fiscal Crisis of the Capitalist State* (New York, St Martin's Press, 1973).
54. 'The Economic Contradictions of Democracy', op. cit. n. 2 above.

6

LABOUR'S DILEMMA: REVISIONISM REVISED

The collapse of the Keynesian, Social Democratic consensus had important consequences for the two parties of government—or rather, the different political tendencies within them. In many respects Labour had been the pacemaker in that settlement while Conservatives accommodated themselves to it. The decline of the consensus is inseparable from changes in the Labour party, notably the widespread sense, in and out of the party, that the dominant or governmentalist wing had failed. This chapter analyses some of the factors which produced this change.

The British Labour Party has long had something of a split personality. From its earliest days it has employed the rhetoric of a party dedicated to the transformation of capitalist society to socialism. The party programme of 1918 lays down as the goal of the party 'the common ownership of the means of production and the best obtainable system of popular administration and control of each industry'. Yet in practice the party, particularly when in government, has had to cope with more mundane problems of here and now, particularly those associated with the country's poor economic performance. In 1931, as a minority government, it rejected schemes of 'tinkering' with a capitalist system in crisis. Robert Skidelsky notes that without a parliamentary majority and, even more important, a theory of transition to socialism the 1929–31 Labour government was in practice doomed to mouthing Utopian slogans and to political impotence.[1] Unemployment soared, international banks lost confidence, and eventually the government collapsed. Some Labour leaders doubted that there could be a parliamentary road to socialism. In government Labour's economic policy had been rigidly orthodox, seeking a balanced budget at all costs. Indeed the Chancellor of the Exchequer, Philip Snowden, and most of his successors, particularly Stafford Cripps, Hugh Gaitskell, James Callaghan, Roy Jenkins, and Denis Healey, have all, sooner or later, proved to be sound money men and at least as economically orthodox as their Conservative counterparts.

The Labour Party in government has regularly been plagued by the nation's economic weakness; ambitious programmes of welfare expenditure and redistribution have had to take second place to policies to boost production and exports and reassure holders of sterling. In 1949 the pound was devalued. In 1966 the Wilson government brought in a severe deflationary package to reassure financial markets and in 1967 it devalued the pound and abandoned its economic plan. In 1974 a minority Labour government took over an economy about to be rocked by the quadrupling of oil prices and, faced by the collapse of the pound in 1976, raised a loan from the International Monetary Fund; the loan was offered, subject to the government making major changes in its economic policy, including cuts in public spending and signing a Letter of Intent to the IMF. The goals of advancing socialism, protecting the interests of trade unions, and helping the smooth working of capitalism in an open economy, have proved difficult to reconcile for the British Labour party. Socialism has often given way to the need to avoid financial crises, reassure business leaders and secure incomes restraint from the trade unions.

Labour and the System

There was little doubt, even before 1914, that the Labour Party had already been integrated into the political system, accepted existing political institutions, and was prepared to work for gradual social and economic change through Parliament. Beliefs in class co-operation and the neutrality of the state were widespread throughout the leadership. The two minority governments of 1924 and 1929–31 had shown no radical intentions, had given no particular favours to organized labour, and were at least as concerned to show their goodwill to finance and business. Ramsay MacDonald's aim was, in his own words, 'to gain the confidence of the country'. The bulk of the parliamentary leadership, in common with that of the TUC, was anxious that the 1926 General Strike should not be successfully used for political purposes. The party also regularly rebuffed approaches in the 1920s for affiliation and even for closer relations by the British Communist party. The Party Conference indeed ruled that no member of the Communist party was eligible for membership of the Labour party. Even in the 'leftist' 1930s the party's leadership kept clear of association with the Communists and far-left ideas. In 1937 it expelled Stafford Cripps and other Labour MPs for advocating a Popular Front.

Any lingering doubts about Labour's constitutionalism were removed by the party's entry to the wartime coalition in 1940. The wartime experience of greater central economic planning and controls by the state over private industry and society helped to make much of the case for socialism. After 1945 the new Labour government enacted an ambitious programme of public ownership. It took over the Bank of England, coal, gas and electricity, each of which had already been the subject of coalition government reports. It also took over rail, road haulage, and, most controversial of all, iron and steel. As noted above on p. 42 the party's 1945 election manifesto argued each case for public ownership on pragmatic grounds, usually to promote more efficiency.

Many on the left of the party regarded the 1945–51 measures as a first instalment towards a much more publicly owned economy. Others on the right, notably Herbert Morrison, argued for consolidation and for demonstrating that existing industries worked efficiently. For the 1950 general election a small list of industries was proposed for future nationalization, including cement, sugar, and water. Labour had a majority of six after the 1950 election and it was clear that another dissolution could not be long delayed. The 1951 election manifesto avoided specific commitments and pledged only that it would consider public ownership when firms were failing to operate in the national interest. Although the party had 13.9 million votes (the highest total ever gained by any British party and more votes than the Tories) it lost the 1951 election and spent the next thirteen years in opposition. New divisions had already appeared in the party over proposals to increase spending on armaments and impose curbs on the

National Health Service expenditure, to add to those on the roll of public ownership. A battle for the post-Attlee succession was also developing between Bevan and Gaitskell. At the Party Conference in 1952 left-wingers took all but one of seven places in elections for the constituency section of the NEC. The party's electoral reverse in the 1955 election gave an added bite to the policy and leadership struggles. On Attlee's retirement in 1955 Gaitskell was elected leader over Aneurin Bevan. Left and right factions quarrelled bitterly over German rearmament, nuclear weapons, and nationalization. Bevan, the obvious leader of the left, finally made his peace with Gaitskell when, as Shadow Foreign Secretary, he spoke against a Party Conference motion in 1957 calling for Britain's abandonment of nuclear weapons.

Revisionism

A number of parliamentarians as well as trade union leaders had long been sceptical of the left's demand for extending public ownership and altering the balance of the mixed economy. The most prominent spokesmen for a revision of the party's thinking in the 1950s were John Strachey, Hugh Gaitskell, and Anthony Crosland. The latter, particularly in his influential *The Future of Socialism*, presented an analysis which was both radical and appealed to the right wing of the party.[2]

Revisionism was a term which was originally applied to the ideas of the German socialist Eduard Bernstein who, at the end of the nineteenth century, noted that the working class was not becoming poorer, society was not being divided into two warring classes, and capitalism was not on the verge of breakdown. Marxist predictions of economic catastrophe, revolution, and a socialist breakthrough were wide of the mark. Hence, Bernstein argued, socialists could and should work through the existing system and introduce reforms, rather than plan for revolution. It is ironic that Bernstein's thinking was influenced at the time by his observation of the success of the Fabians in Britain. Some sixty years later British revisionists wanted the Labour party to emulate the German Social Democrats who, at Bad Godesberg in 1959, abandoned their Marxist ideology.

Crosland similarly reminded his readers that capitalism was not on the point of collapse. The separation in many firms of the ownership of shares from managerial authority and the blurring of lines between public and private industry meant that 'capitalism was reformed almost out of recognition'.[3] Crosland claimed that managers had become more socially responsible, as much interested in creating a stable economic environment as in the pursuit of profits, and that the Government was able to influence firms without directly controlling them. Moreover, the Government in combination with trade unions was now able to provide a countervailing power to business. Crosland pointed to the many post-war gains the left had made; witness the emergence of the welfare state, managed economy, and full employment, all of which had been accepted by the Conservatives and achieved in a mixed economy. There were no votes for the Tories in a return to the 1930s; therefore, 'the national shift to the Left may be accepted as permanent'.[4]

The party became less ambitious in its plans for public ownership. The 1955 manifesto merely promised to renationalize iron and steel and road haulage and take over sections of chemicals and machine tools. A party statement in 1957, *Industry and Society*, said, 'the Labour party recognises that . . . large firms are as a whole serving the nation well'. In the 1959 manifesto the renationalization proposals remained (for largely historical and sentimental reasons), together with an assurance (or threat) that 'We have no other plans for further nationalisation. But where an industry is shown . . . to be failing the nation we reserve the right to take all or any part of it into public ownership if this is necessary.'

For Crosland the question of the ownership of industry was no longer very important; it could not be the litmus test of socialism. The government could still plan the economy by means of the budget, investment incentives, and controls on the location of industry. The new goal of socialism should be equality, particularly equality of opportunity, which could be achieved largely through programmes of public spending and narrowing unearned differences in incomes. Crosland argued for the importance of economic growth if redistribution and greater equality were to be achieved without conflict.

Revisionism was also interested in and affected by the growth of studies in electoral sociology; indeed, it was a theory of how to win elections in an age of growing affluence and declining social class differences. Academics were speculating about the end of ideology and how, under the impact of affluence and full employment, the working class had lost its socialist zeal. A number of right-wingers in the PLP had, for various reasons, long wanted to tone down the party's sectional appeal to the working class and Clause 4 socialism. The 1959 election defeat, the third in succession, was a key event for such people. In that election the party's share of the vote (43.7 per cent) was its worst since 1935. Many commentators linked Labour's steady loss of votes with the decline in the size of the manual working class and of heavy manufacturing industry and predicted a continued Labour decline in the future. Those who argued that the party had failed to adapt to a more affluent society were matched by those who argued that the party needed to be bolder in promoting socialist policies. The *Must Labour Lose?* survey by Abrams and Rose was significant in this debate.[5] It showed that trade unions and public ownership were unpopular with the electors, and that Labour was widely seen as being old-fashioned, sectional, too tied to the working class, and having weak leadership. Much of the right-wing leadership of the party respected these findings, not least because it confirmed many of their own views.

Hugh Gaitskell took the initiative in a special two-day Party Conference in November 1959. Inevitably the Conference was dominated by the election defeat. Gaitskell suggested that changes in the composition of the working class, the onset of full employment, the spread of affluence and house-ownership, and Conservative charges that Labour proposed to nationalize most industries had damaged the party. Public ownership was a means not an end. Clause 4, he suggested, was incomplete as a statement of the party's aims; as expressed it encouraged the belief that the party wanted to nationalize everything, which it did not. His proposal to add a further set of party aims as an amendment to Clause 4 met fierce opposition, not just from the left but from many trade unions. Clause 4 was a symbol of the party's labourism and any attempt to dilute it aroused resistance. Harold Wilson compared it to trying to take Genesis out of the Bible. In the end Gaitskell had to give way partly because he knew that another battle, on defence, would be developing in the party. He had to settle for a twelve-point statement of the party's aims as an amendment to the constitution. Stephen Haseler, in his study *The Gaitskellites*, notes that this was one of the first occasions when the major unions refused to back the parliamentary leadership.[6] On most home policy issues the political left took the initiative, usually proposing a shopping list of industries for nationalization, and the major trade unions could be relied on to line up against them. But the revision of Clause 4 was an initiative by the leadership, an attack on the most traditional part of Labour's ideology, and the trade unions acted on behalf of the status quo.

Conference Politics

In 1960 Conference defeated the platform on unilateral disarmament and opposed the stationing of American missiles in Britain. Gaitskell defied Conference and in doing so created something of a constitutional crisis for the party: who had final authority in the party when the PLP disagreed with Conference? In the past the party leadership, usually in alliance with the large trade unions, had 'managed' Conference so that it would win the votes on the major issues. Although Gaitskell persuaded Conference to reverse the defence decision in 1961, he paid tribute to its authority by campaigning for a change of policy. Interestingly, the authority of Conference had not been raised in earlier years when it had supported the parliamentary leadership. After the death of Gaitskell in 1963 Harold Wilson was elected leader over George Brown on the second ballot. With another Labour government installed in 1964, and Conference able to judge by deeds, relations between it and the party leaders worsened again after 1966. Between 1966 and 1969 Conference defeated the leadership over pit closures, Vietnam, prescription charges, handling of the economy, and incomes policy. In contrast defeats had been extremely rare between 1945 and 1964. The negative votes were ignored by the government and Mr Wilson's reply was usually to the effect that the Labour government would continue to govern in spite of what Conference said.

Thus revisionism seemed to be triumphant in the party leadership. As Lewis Minkin notes in his study of Conference, the authority of the latter was almost extinguished, as the government went its own way.[7] The vocabulary of ministers increasingly referred to the need for increased investment in industry, greater productivity, more exports to help the balance of payments, and for the requirements of production to be placed ahead of those of social expenditure. In economic policy the 1966 government showed itself to be rigidly orthodox and David Howell claims that 'during the Sixties the party's official appeal was more purely national, practical, and governmental than ever before'.[8] Mr Wilson more than once boasted that Labour, so long a party of protest, was now a party of government. Many aspects of the government's record disillusioned the activists. The collisions with the trade unions over incomes restraint and *In Place of Strife*, the slow rise in living standards, reports by reputable groups about the shortcomings of the government's welfare record, and ministers' cavalier treatment of Conference provided many targets for critics.

Public ownership was also relegated in importance, although iron and steel and road haulage were renationalized. In 1963 Mr Wilson had redefined socialism in terms of the scientific revolution, a concept that had little to do with public ownership.

> Socialism, as I understand it, means applying a sense of purpose to our national life: economic purpose, social purpose and moral purpose. Purpose means technical skill . . .

The 'normal' post-war pattern of power in the Labour party was for the parliamentary leadership to be supported at Conference by the major trade unions, particularly the 'Big Three', the Transport and General, National Union of Mineworkers, and General and Municipal Workers' Union. In the 1950s there was a substantial degree of oligarchy in these trade unions. Historically, as McKenzie points out, the major unions had acted as a 'Praetorian Guard' for a right-wing parliamentary leadership against the left in the PLP and in the constituencies.[9] In reality, however, the pattern was a system of alliances, involving the bulk of the trade unions, most of the PLP, and some constituencies against leftist minorities in the first two groups, and most constituency delegates.[10]

The support of trade union leaders for the parliamentary leaders rested in part upon an assumption of there being a separation of spheres between the political and the industrial, which were left to the parliamentary and trade union leaderships respectively. At the 1947 Conference the Transport and General Workers' Union leader, Arthur Deakin, had bluntly warned the politicians, 'the questions of wages and conditions are questions for the trade unions.' But after 1964, with Labour in office again and determined to modernize industry, promote economic growth, and slow down inflation, policies for incomes and industrial relations were placed on the agenda. In 1965 the government introduced a bill to give it power to delay wage settlements while reference was made to a newly established national Board for Prices and Incomes. This was lost because of the 1966 general election but reintroduced in the new Parliament and followed by a wage freeze in July. There followed three years of statutory incomes policy. They strained the Labour alliance and the tension reached its highest point with the publication in January 1969 of the Labour government's plans *In Place of Strife* and the historic decision to legislate in certain situations for penal sanctions against trade unions. In the event sustained opposition by the unions and by MPs forced Mr Wilson to abandon his plans.

These events raised difficulties for running the party. The trade unions became more politicized, partly as a reaction to government policy and partly because of an infusion of left-wing activists. In the late 1960s the two biggest unions, the Transport and General (TGWU), and the Engineering Union, had left-wing general secretaries. But more significant was the trend for general secretaries of the unions not to be in secure control of their delegations. Although the unions were still able to dominate the annual Party Conference, the general secretaries were less able to use their votes on behalf of the parliamentary leadership.

The inconclusive nature of the revisionist victory was seen after 1970. As Minkin notes, given the rather bogus nature of many of the votes cast at Conference, it was probably an illusion to think that the grass roots membership had ever accepted a revisionist policy. He adds,

> the victory of revisionism was achieved by a combination of procedural manipulation and the winning of strategically placed supporters rather than by mass persuasion of the majority of the rank and file of the unions and the party. Its basis amongst party and union activists had changed little in the previous fifteen years.[11]

The 1970 election defeat and subsequent attacks on the Labour government's record weakened the parliamentary leadership. The authority of Conference was reasserted. Mr Wilson was soon a lone defender of the record of his administration (as Mr Callaghan was to be after 1979). The party's shift to the left during the 1970s was one which occurred on a number of fronts.

Revisionism Halted

After 1970 the trade unions were determined to protect themselves against further legislation under a Labour government on incomes and industrial relations. But they also wanted a forum for co-operation. A Liaison Committee between the unions, NEC, and PLP was established in 1972 and it worked out a so-called Social Contract, embodied in the statement 'Economic Policy and the Cost of Living' in February 1973. This stated that a future Labour government would repeal the Industrial Relations and Housing Finance Acts, provide new rights for the trade unions, impose a system of price controls, increase public spending on pensions and welfare programmes, and redistribute wealth and income through a progressive taxation policy and social spending.

The party outside Parliament continued to move to the left after 1970. The decline of revisionism on policy questions after 1970 was reflected in Conference decisions between 1970 and 1973 which included a list of new commitments to public ownership. It was also reflected in *Labour's Programme for Britain, 1973* which promised a fundamental and irreversible shift in the distribution of wealth to working people and their families. It also proposed to set up a state holding company which would acquire a major stake in manufacturing industry. To tackle the concentration of economic power in the hands of large companies and multinational corporations the government would be empowered to make planning agreements with a hundred leading manufacturing firms. An Industry Act would give it power to intervene in firms, secure information, and even purchase firms. The manifesto for the February 1974 election promised to take into public ownership North Sea oil, the docks, aircraft, and shipbuilding. Finally, public ownership would not be confined to loss-making firms or sectors but extend to profitable firms, 'where a public holding is essential to enable the Government to control prices, stimulate investment, encourage exports, create employment, protect workers and consumers from the activities of irresponsible multi-national companies'. All this was a far cry from the views of Crosland in the 1950s.[12]

Opposition to the new Conservative government was particularly bitter over housing finance and industrial relations, and much of the battle was fought outside Parliament. There were trade union strikes against the Industrial Relations Act, which was eventually made unworkable when most trade unions refused to register. The left-wing local authority in Clay Cross defied the provisions of the Housing Finance Act. The National Union of Mineworkers successfully challenged the government's incomes policies in two bitter strikes in 1972 and 1974. In Parliament the Labour party was badly split over the government's application to join the European Community. The Party Conference in October 1971 came out against the terms of entry negotiated by the Conservative government but sixty-nine Labour members defied a three-line whip and voted to support the government's terms of entry. In 1972 Roy Jenkins resigned his position as Deputy Leader when the party came out in favour of holding a referendum on the issue. The right in the political party was weakened by the European issue, because it was so closely associated with a cause that was unpopular in the party and the country.

Labour was returned to office as a minority government in February 1974, and again with a wafer-thin majority in October. The trade unions achieved the repeal of the hated

Industrial Relations Act and further amendments to trade union laws in the Trade Union Labour and Relations Act (1976) and the Employment Protection Act. Because of widespread union fears about mounting inflation and the failure of the Social Contract to persuade bargainers to exercise moderation in wage settlements the union leaders took the initiative in setting up a non-statutory incomes policy in 1975 which lasted, after a fashion, until 1978. Contacts between leading members of the Trades Union Congress and the Labour government were maintained in various ways. The TUC Labour Party Liaison Committee met several times a year to discuss current issues; it consisted of six Cabinet Ministers, nine members of the NEC, and six representatives of the TUC. It renewed the Social Contract each year, covering a wide range of government policies. A much heralded National Enterprise Board, which had been intended to make planning agreements with firms and extend government intervention throughout the economy, was also established. In the end, however, only one planning agreement was achieved, with the ailing Chrysler car firm, and the industrial strategy did little more than provide subsidies for loss-makers. Deflationary economic policies and minute improvements in living standards made the government deeply unpopular among its own supporters. As long as the TUC was prepared to acquiesce in incomes policy and in public-spending cuts after 1976, however, the left-wing-dominated NEC was outflanked. The breakdown occurred in 1978 when Mr Callaghan proposed a 5 per cent limit for incomes in the next round of pay policy. This was rejected by both the TUC Conference and the Labour Party Conference. The government tried to make the policy stick but it collapsed in the Winter of Discontent (see above p. 130).

The Labour party's position in the country swiftly crumbled. Between January and February 1979 Gallup showed the Conservative lead over Labour increasing from 7½ per cent to 20 per cent and in March it was still 14½ per cent. Labour's once commanding position on key issues also worsened. Between August 1978 and February 1979 MORI found that Labour's nine-point lead over Conservatives for having the best policy on prices/inflation fell to a deficit of twelve points.

On handling industrial relations its nine-point lead was transformed to a thirteen-point deficit. There can be no doubt that the inability of a Labour government to control its 'natural' allies dealt a fatal blow to the 'social partners' approach, particularly since the opposition party had already abandoned it.

Party Versus Government

A key event in the history of the Labour governments was the International Monetary Fund (IMF) rescue package for the pound in 1976. Investors were steadily selling sterling and at one point the pound fell to $1.57. Britain was approaching the stage where the government might not be able to repay debts. The government arranged massive loans from the IMF in return for cuts in public spending, controls on the money supply, and a reduction in the PSBR. Inflation was indeed reduced (to 8.2 per cent at the end of 1978) and the pound recovered, but Labour's plans for ambitious social reforms had to be postponed. The collapse of the pound and the negotiations with the IMF were yet another reminder of the weaknesses of the British economy and of the limited power of governments.[13]

The National Executive also moved to the left. In the period from 1974 all but one of the seven constituency members were representatives of the left. In 1976 Denis Healey, Chancellor of the Exchequer, was voted off it and in 1977 and 1978 over 80 per cent of the votes cast by the constituency parties went to left-wing candidates. Four of the five members in the women's section of the NEC were also on the left. The twelve members of the trade union section were harder to classify but could not counterbalance the left majority. The government had reliable support from no more than ten of the twenty-nine members of the National Executive. The referendum on whether Britain should remain in the European Community exposed the conflict. The NEC voted by eighteen votes to eleven to recommend that the Labour party special conference in April 1975 reject the terms negotiated by the Labour government and Conference endorsed the NEC view by a two-to-one majority. It took some deft footwork by the leadership to prevent the NEC actually committing the government party to campaign against the NEC. The NEC later opposed the government's decision to support direct elections to the European Parliament. It also opposed the government's expenditure cuts in 1976 and the later stages of incomes policy. Between 1945 and 1969 the platform had rarely lost a vote at Conference; between 1970 and 1979 it lost thirty-two.

At the same time the NEC resisted investigations of alleged left-wing infiltrations of the party. It decided not to publish a report on the Trotskyite entryism which had been prepared by party officials. It also endorsed the choice of a declared Trotskyite as party youth officer. In cases where constituency parties were reluctant to renominate the MP (the most famous case being the refusal of Newham North East to renominate Reg Prentice, a Cabinet member and leading right-winger who subsequently crossed the floor and became a junior minister in Mrs Thatcher's first government), right-wingers thought the NEC was at least acquiescing in left-wing activist pressure on MPs. In 1973 the party abolished its list of 'proscribed' organizations (usually Communist and far-left groups), allowing many Communist 'front' groups to have closer relations with the party.

The growing gap between the policies of the Labour government and the party outside grew more apparent. Conference and NEC regularly advocated policies of greater state ownership and higher public expenditure, all contrary to the government's economic policies. The contents of the NEC's *Labour Programme* 1976 were compiled from Conference resolutions and were an official statement of the party's medium-term strategy. This proposed nationalization of banks and insurance companies, compulsory planning agreements, and a wide range of welfare benefits and was overwhelmingly endorsed by Conference. The policies were far to the left of those in the 1950s and 1960s. The *Programme* contained no commitment to the mixed economy, was hostile to the EEC, and was distinctly neutralist, even anti-NATO.

Yet at the same time ministers were trying to adhere to the guidelines of the agreement with the IMF and restore the confidence of industry and holders of sterling. In government, there was little support for the left's Alternative Economic Strategy—a 35-hour week, import controls, withdrawal from the European Community, price controls, and greater use of public ownership and planning agreements with major firms.

The Left's Critique

How does one explain the disappointment at the records of the Wilson and Callaghan governments? After both periods of government the cry of 'betrayal' was raised, usually by groups on the left, who charged that crucial Conference resolutions had been disregarded and manifesto pledges broken. For critics on the left as well as on the right the record seemed to be part of a pattern too familiar to be explained by chance, the lack of a parliamentary majority (between 1977 and 1979), the personalities of Wilson or Callaghan, or economic circumstances. The pattern was one of 'limited success and frustration of radical expectations'. Viewed negatively, the record of the 1966 government included cuts in living standards, statutory controls on incomes, confrontation with the trade unions, and a record of economic growth that was, to quote Crosland, 'lamentable'. The 1974 government again ended with the two sides of the Labour movement in conflict. There was no promised wealth tax or industrial democracy and the National Enterprise Board was hamstrung. Again, there had been pay controls and spending cuts, and the influence of regular Conference votes against government policies was minimal.

Labour governments in the 1960s and 1970s had shown themselves to be at least as economically orthodox—placing the strength of sterling or the balance of payments ahead of production or social spending—as any Conservative government. Ministers claimed that they had got the acceptance of the unions to wage restraint and austerity in the economic 'typhoon' of 1974–6 (Michael Foot's term) and delivered the consent of the unions to a fall in living standards. A typical left-wing view was that Labour's efforts to demonstrate its national outlook and its fitness for government 'in practice has involved the subordination of working-class interests to the imperatives of capitalist accumulation'.[14]

To those on the left who looked for more sweeping socialist economic transformation the failure had occurred because of what Miliband has termed gradualism or labourism; a belief that socialism could come via politics as usual, reliance on parliamentary methods, and a gradual extension of public control over the economy. For Miliband, Labour's history since 1931 has been one of 'MacDonaldism without MacDonald'.[15] But it could also be argued that any British government was unable to shape the policy of powerful multinational corporations and was indeed the weakest part of an international capitalist system which was undergoing a general crisis. International capital wanted to keep wages low for the sake of competitiveness, did not welcome strong trade unions, and sought low inflation. The dilemma for a socialist government in western capitalist society was that aspirations for full employment, redistribution, strong trade unions, a well-funded welfare state, economic planning, and accountable economic power were policies which would frighten capitalists. If this analysis as developed by David Coates has merit then there was and is no point in looking to a Labour government for a socialist breakthrough, partly because it is wedded to parliamentary methods and the mixed economy. The more revolutionary strategy favoured by Coates would have to come from a different Labour party or from another party.[16] But perhaps there was little point in blaming Labour if it was caught up in a crisis of international capitalism. As the Socialists in France found after 1981 there was not much that a socialist government could do acting alone.

These developments posed a problem for revisionists. Crosland had always argued that greater social spending and redistribution depended on economic growth. Without growth, taxes would increase and place pressures on take-home pay; in turn this would generate demands for wage increases and, in the absence of economic growth, inflation. But the Wilson governments in the 1960s had abandoned the pursuit of economic growth, had made the balance of payments its top priority, and had brought many low-paid workers into the tax net. In the 1960s support for spending on social services dropped most sharply among those who were disappointed economically by the 1964 Labour government.[17] Affluence was required to make people feel sufficiently well off to support 'altruistic' social policies.[18] There was also growing evidence in the 1970s that increases in the so-called social wage did not dampen pressure to maintain or increase wages and salaries. Crosland, indeed, in his last work was no longer so confident that expanding public spending was the route for egalitarians. He noted that much spending actually went to middle-class interests (for example in higher education, commuter trains, and occupational and tax relief for pensions and mortgage interest on homes), and was therefore prepared to give priority to public spending which was directed more to the working class. As he commented:

> In the social services, much of the spending has gone on creating large bureaucracies of middle class professional people. Where we once were sure that better education would enable working class children to catch up with the children of the middle classes, we now know, thanks to the work of Jencks and his associates in the United States—that . . . the character of the schools output depends largely on a single input, namely the character of its entering children, everything else—the school, the school budget, its policies, the characteristics of the teachers—is either secondary or completely irrelevant.[19]

In the negotiations for the IMF loan Crosland led a group of ministers in resisting the Treasury's demands for spending cuts. But in the end they gave in— 'It's nonsense but we must support it' (Crosland)—to buttress the authority of the Prime Minister and Chancellor. By the time of his death in 1977 it was already clear that Crosland's optimistic expectations had gone wrong, particularly about economic growth, redistribution, the value of public spending, and the success of capitalism.[20]

The most articulate spokesman for a change of direction was Tony Benn, the hero of the constituency parties. In bold language at annual Conferences he dissociated himself from many of the policies of the government of which he was a member. He was to campaign more vigorously for changes in policy and in the party's constitution after the 1979 election defeat.

Electoral Decline

Since the war the Labour party had been perhaps the most united working-class party in any Western European country.

TABLE 6.1 *Decline of Labour Working-Class Vote (or percentage of popular vote)*

Votes	1959 Non-manual	1959 Manual	1964 Non-manual	1964 Manual	1966 Non-manual	1966 Manual
Conservative	69	34	62	28	60	25
Liberal or Minor Party	8	4	16	8	14	6
Labour	22	62	22	64	26	69

Votes	1970 Non-manual	1970 Manual	Feb. 1974 Non-manual	Feb. 1974 Manual	Oct. 1974 Non-manual	Oct. 1974 Manual
Conservative	64	33	53	24	51	24
Liberal or Minor Party	11	9	25	19	24	20
Labour	25	58	22	57	25	57

Votes	1979 Non-manual	1979 Manual	1983 Non-manual	1983 Manual
Conservative	60	35	58	33
Liberal or Minor Party	17	15	26	29
Labour	23	50	17	38

Source: I. Crewe, 'The Electorate: Partisan Dealignment Ten Years On', *West European Politics* (1984), p. 194, and sources cited there.

The trade union movement was not split politically or industrially and there was no rival socialist party or strong Communist party competing for working class votes. Moreover it was usually the only socialist party in a position to form a government on its own because it was unencumbered by operating in a multi-party system or a proportional electoral system. Yet by 1979 Labour had undergone the most spectacular electoral decline of any socialist party in Western Europe. It lost votes at every general election bar one between 1951 and 1979 and its vote tumbled from 47 per cent in 1966 to 37 per cent in 1979.

The erosion of Labour support among the working class is seen in Table 6.1; between 1964 and 1979 it fell from 64 to 50 per cent. At the same time its lead over the Conservative party among this class fell from 36 to 15 per cent. There was a growing gap between the policy preferences of Labour supporters and the party's programme. In 1964 many Labour identifiers supported the party's position on three important tests of socialism: more public ownership, more spending on social services, and not accepting that the trade unions had too much power. By 1979 only a third of Labour identifiers supported these policies (Table 6.2), and manual workers were more in sympathy with many Conservative rather than

TABLE 6.2 *Long-Term Trends in Support for Labour Party Policies Amongst Labour Supporters*

	1964	1966	1970	1974	1979	Change 1964–79
Percentage of Labour identifiers:						
In favour of nationalizing more industries	57	52	39	53	32	−25%
Who do not believe that trade unions have 'too much power'	59	45	40	42	36	−23%
In favour of spending more on the social services	89	66	60	61	n.a.	−28% (1964–Feb 1974)

Source: 1964, 1966, 1970 Butler and Stokes Election Surveys; Feb. 1974, Oct. 1974, 1979, British Election Study cross-section surveys.

Labour proposals; even on issues on which the party was united, as on restoring trade union rights, stopping council house sales, extending nationalization, comprehensive education, and providing more social service benefits, working class voters were divided.[21]

There was also change and decline among the grass roots. If we discount the trade union affiliations (of whom many are reluctant or involuntary members, and indeed may not vote for the Labour party), individual membership fell from over a million in 1952 to less than 300,000 in 1979. The fall in party membership seemed to be part of a general trend: with the growth of television, new forms of recreation, and greater mobility, people seemed to turn away from organized political parties. Yet this decline occurred at a time when membership in other voluntary organizations was actually increasing. Some of the smaller Labour parties in city centres were easy targets for 'takeovers' by left-wing activists.

There was a significant radicalization of the Labour party in local government in a number of cities. The heavy losses suffered by Labour in the local elections in 1968 and 1969 swept away many senior Labour councillors. In places like Manchester, Liverpool, Sheffield, and many London boroughs they were replaced by younger, more assertive, and more left-wing councillors. Many of the councillors were employed in neighbouring local authorities, or were officials in local public-sector trade unions, and were in many respects a public-sector political class.[22] After 1979 they were frequently in conflict with the Conservative government because of their 'high' spending and 'high' rates. They were local socialists and Keynesians, spending money in part to regenerate their local economics, while the central government was pursuing a very different policy.

Although we have no long-term evidence on the point, it may be that the smaller Labour membership was both more committed and more left wing. A survey of Labour Party Conference delegates by Gordon and Whiteley found that the majority of delegates were white-collar, though self-described working class, and employed in the public sector (Table 6.3), just as the parliamentarians were increasingly middle class.[23] By large margins they supported Tony Benn for the leader-

TABLE 6.3 *1978 Labour Conference Delegates*

Public sector	60
Skilled workers	22
White-collar workers	70
Among white-collar workers:	
self-described working class	72
self-described middle-class	22
objectively clearly middle-class	57

Source: P. Whiteley and I. Gordon, *New Statesman*, 11 January 1980.

ship of the party as well as a whole range of left-wing policies and vigorously denounced the leadership for 'betrayal'. On both counts they were markedly out of line with the views of most Labour voters.

Within the PLP the left wing continued to make progress, even though it was weaker there than in the NEC or Conference. Right-wing MPs formed a Manifesto Group in 1974 and then a Campaign for Labour Victory (CLV) in 1977. Some of the MPs who were to lead the breakaway to the SDP were prominent in the CLV. Within Parliament there was also the emergence of a new 'hard' left, which was much less respectful of parliamentary traditions. The replacement of sitting MPs over time through retirement and reselection due to constituency redistributions was helping the left. Perhaps the culmination of this was the election of Michael Foot as leader of the party over Mr Healey in 1980.

Party Reforms

By 1979 a number of critics were pointing to a cycle in the history of the Labour party. In opposition, the extra-parliamentary bodies, particularly Conference, became more assertive and influential, the policies often shifted to the left and the party leader was no longer the Prime Minister, with authority and patronage. But in government the leaders were more concerned with policies which they judged to be politically acceptable, administratively practicable, and affordable. Many of the policies of the government brought them into conflict with Conference and the NEC. As members of a 'national' government and with responsibilities to the electorate, most ministers were less orientated to the party membership. Mr Benn was the most articulate spokesman for the thesis of 'betrayal' of the movement by its parliamentary leaders.

> We have seen twenty years of surrender. Since 1959 the Parliamentary leadership of the Labour Party has been going along with the idea that the post-war consensus built upon full employment and the welfare state was a permanent feature of life in Britain and that trade unionism would be brought into a position where it helped to run it. That response has failed to command the support of our people because they have seen first that it did not contain within it any element whatsoever of transformation and second that even by its own criteria it failed. That policy could not bring about growth and could not extend freedom.[24]

The lesson which the activists learnt from this recurring pattern was that there appeared to be little point in winning policy battles in opposition if Labour governments were going to dilute them in office. Spearheaded by the Campaign for Labour Party Democracy and the Labour Co-ordinating Committee, the activists decided to change the structure of the party to ensure that party policies were carried out and that the parliamentary party was made subordinate to Conference. The call for greater party democracy was mounted by the left wing to make the PLP more accountable to Conference and MPs more accountable to local party activists. Shifting the balance of power in the party became the left's preferred way of closing the gap between Conference and PLP and preventing any alleged 'betrayal' by the latter. It was also a way of closing the gap between myth and reality of which McKenzie had made so much.

Here we have to take account of the power of a myth as a goal to give rise to patterns of behaviour which go some way to fulfilling that myth. Revisionism in the Labour party in the 1960s was a theory of economic management and electoral strategy, and became also a defence of the Labour MPs' independence from instructions by Conference. Revisionists argued that modern managed capitalism had changed and that it could provide full employment and economic growth.

Its success rendered irrelevant questions of public ownership, and economic growth provided funds for the welfare state and made redistribution easier. But what if successive Labour governments, while ignoring Conference, are widely thought to be failures, particularly in improving living standards and promoting economic equality and other goals? What if they fail to win elections, though defiance of Conference has been justified with a view to cultivating the electorate? What if they divide the party (for example, *In Place of Strife* or incomes policies) on the grounds that such policies are economically and politically necessary? In fact, as Labour governments failed to get economic growth, imposed wage restraints, and quarrelled with the unions, so the influence of the left in the extra-parliamentary party grew, as did that of the extra-parliamentary party.

The Labour party has always faced a dilemma because if its constitution's acknowledgement of the sovereignty of the party Conference were taken literally, then MPs would be merely delegates of the party Conference. Some commentators have argued that the party's idea of intra-party democracy is incompatible with the ideas and practice of representative democracy and parliamentary sovereignty.[25] If the Party Conference really did 'instruct' Labour MPs then how could the latter be accountable to the voters? Where there were differences in policy choices between Labour voters and Conference, then which group should the MP represent? But by tradition and usage the role of Conference had been merely advisory. Tony Benn claimed that giving greater power to the party activists and to the annual Conference would make the leaders and the MPs more responsive to the movement. Party democracy was a means of ensuring the implementation of socialism. The groups circulated model resolutions, had them adopted by sympathetic constituency parties and trade union branches, and successfully exploited the party's tradition of rank and file participation and the myth of Conference sovereignty.

Demands for constitutional reform centred on three features: NEC control of the manifesto, mandatory reselection of MPs within the lifetime of Parliament, and election of the leader by party members. At the 1979 Conference in Brighton a proposal to introduce mandatory reselection was carried, as was a proposal to give the NEC control of the manifesto. The same Conference, however, defeated a proposal to allow the mass membership to elect the party leader. The NEC, under trade union pressure, then appointed a fourteen-member Commission of Inquiry with representation from the NEC, trade unions, and the PLP. Its proposals for constitutional reform, including an electoral college to choose the party leader, were reviewed at a meeting held at Bishop's Stortford in June 1980. Jim Callaghan and Michael Foot had been briefed by the PLP to retain its role in the election of the leader and in writing the manifesto. However they failed to block the idea of the electoral college, a concession which shocked many MPs. Potential defectors like David Owen and Bill Rodgers saw the outcome of the Conference as a sign of the parliamentary leadership's lack of fight. But Mr Callaghan and Mr Foot accepted the college partly because it was the best thing on offer and partly because they thought it might collapse under its own complexity.

The proposals were submitted to the 1980 Conference at Blackpool. Votes for mandatory reselection and the election of the party leader by some form of electoral college were carried but the proposal to give the NEC control of the manifesto was lost. A further Conference was held at Wembley in January 1981 to decide on the details of how the party leader would be elected. The eventual resolution granted the trade unions 40 per cent of the vote, MPs 25 per cent, and constituencies 25 per cent. The reforms primarily concerned the relationships between different party institutions and had been promoted and resisted by the left and right wings as a means of promoting their respective ideologies and interests. If there had been a greater agreement in the party on policy then the debate would have been less divisive. Many other issues clearly related to democracy in the party, such as the block vote of the trade unions, one member one vote elections, and the system by which trade unions and constituency parties actually gained their votes at Conference, were ignored.

When Mr Callaghan announced his resignation as party leader in October 1980, Denis Healey was the clear favourite to succeed him. Michael Foot was Deputy Leader but, because of his age (67) and his assumed support for Peter Shore, he was not regarded as a likely winner. In the end, influenced by heavy pressure from trade union leaders, constituency parties, and his own assessment that he would do better than any other left-winger against Mr Healey, he stood. Peter Shore and John Silkin were eliminated on the first ballot and Mr Healey ran ahead of Mr Foot. But on the second ballot Mr Foot won by 139 votes to 129. His election was widely regarded as a vote for reconciliation and unity—he certainly gained a handful of votes from MPs on the centre and right who thought that he offered the best chance of avoiding a split in the party, as well as some who hoped that his election would hasten a split. He was the first party leader since Lansbury in 1933 to come clearly from the left.

The new machinery for electing leaders was tested fully in the election for the deputy leadership in 1981. The contest was another blow to the cause of party unity. In 1980 Denis Healey had been elected unopposed as Deputy Leader by MPs under the old system, he was now opposed by Mr Benn and Mr Silkin. Mr Foot appealed on a number of occasions for Mr Benn to withdraw in the interest of party unity, and indeed challenged Mr Benn to run against him for the leadership on the grounds that his accusation of 'betrayal' by the PLP was an implicit attack on Mr Foot himself. In spite of his attacks on Mr Benn, however, Mr Foot did not provide the endorsement for Healey that the latter's supporters were hoping for. The election campaign was a drawn-out and bitterly contested affair. When the votes were finally cast Mr Silkin was eliminated on the first ballot and Mr Healey eventually defeated Mr Benn by a mere 0.8 per cent of the electoral college vote. His lead among MPs (137 to 71) and trade unions (3.9 million to 2.3 million votes) just overcame the 4 to 1 support of the constituency parties for Benn. A number of the old or 'soft' left who supported John Silkin on the first ballot for the Deputy Leadership refused to vote for Tony Benn on the second round. To the fury of the latter's supporters 37 MPs abstained, including Mr Kinnock. The election had been divisive, and the minimal consultation which some trade unions had with their members and, in the case of the Transport and General Workers' Union, the delegate's rejection of the rank and file verdict brought no credit to the new procedures.

The election climaxed an unhappy period for the Labour party. There was intense divisiveness and unhappiness in many constituency parties, as left-wing activists pushed their cause. Fear of deselection certainly led a number of MPs to desert to the Social Democratic Party, but members of the 'soft left' were also under pressure from activists. In the end, however, only eight Labour MPs were lost because of reselection; none were on the left and all of the replacements were on the left. Party Conferences were trying occasions for the leadership as leaders and platform speakers were regularly shouted down by the floor. David Marquand had been a Labour MP from 1964 until 1976 when he left to join the EEC Commission (he eventually joined the SDP). He thought that the party had not only lost touch with many working-class supporters but was also more intolerant of middle-class intellectuals. He called this the 'cult of the tea room':

> The central tenets of the 'cult of the tea room' are that the Labour Party is, or at any rate ought to be, not merely a predominantly but an exclusively working class party: that the working class can be properly represented only by people of working class origin, who alone understand its aspirations and have its interests at heart: that middle class recruits to the party, so far from being assets, are liabilities, who have no rightful business in it and who, if they do manage to join it, ought at least to keep their origins dark: and that the elaborate intellectual constructs of the middle class Radical are therefore, at best, unnecessary, since the party can be guided much more satisfactorily by the gut reactions of its working class members, and at worst positively dangerous, since they may lead the party away from its working class roots.[26]

After 1979 the party quickly abandoned many policies the last government had followed, just as happened after 1970. By 1981 Conference favoured unilateral disarmament, closure of all nuclear bases in Britain, abolition of the House of Lords, massive increases in public spending, and withdrawal from the EEC. It also adopted the thirty-five-hour week and import controls. Right-wing revisionists were again a target of party criticism. Indeed, the major defenders of the record of the government were people like David Owen, Shirley Williams, and Bill Rodgers who, ironically, were soon to found the Social Democratic party. The Labour party also paid a heavy price in loss of electoral support. In January 1981 at the time of the special Party Conference, Labour (with 46 per cent support) enjoyed a commanding lead of 11 per cent in the Gallup poll. This disappeared by February and at the end of the year the party had sunk to its record low in Gallup, 23.5 per cent. The rise of the Alliance (which made by-elections a trial for Labour) and then the Falklands limited any chance of an electoral comeback by 1983. Mr Foot was an electoral liability and his poll ratings for a party leader were an all-time low.

At the 1981 Conference the right wing was at last helped by the trade union block vote to purge the NEC. It managed to sweep off five left-wingers and to replace them by five right-wingers and the advance continued in 1982. The Labour party therefore presented something of a paradox by 1982: its leader and its policies were the most left wing the party had had in the post-war period and its constitution had been reformed along the lines favoured by Tony Benn and the left. But the NEC had been recaptured by the right, the Shadow Cabinet was largely right wing (only Alfred Booth had voted for Tony Benn on the first round of the deputy leadership election), and the party was on its way to purging the current symbol of left-wing activism, the Militant Tendency.

Militant is a Trotskyite organization formed in 1950 and dedicated to revolutionary socialism. The different Trotskyite groups had minuscule popular support and the Workers' Revolutionary Party fared disastrously in the 1979 general election, winning a mere 0.5 per cent of the vote in the sixty seats for which they stood. Militant, however, had decided to work within the Labour party. Ostensibly, it merely sold its newspaper, but it also managed to infiltrate the party's youth organization and a number of constituency parties, particularly in Inner London, Bradford, and Liverpool (four of the seven parliamentary candidates in Liverpool by 1983 were supporters of Militant). In 1977 the NEC had resisted investigation of alleged Trotskyite infiltration of the party.

By 1981, however, a more right-wing NEC was determined to take action to root out the Tendency. It voted to accept a new report by the party's national agent which argued that Militant, as then organized, should no longer be tolerated within the party. The NEC proposed to compile a register of those non-affiliated groups of members which could apply for official recognition and be allowed to operate within the party. But such groups had to have 'an open democratic structure' and provide details of their finance, membership, and minutes. The Party Conference in 1982 carried the motion, largely by the trade union block vote (by 5 million to 1.5 million votes). However, the process of identifying and rooting out supporters of the Tendency was to be left to the constituency parties, some of whose activists were opposed to a witch hunt. Although the NEC eventually expelled five members of the paper's editorial board, militant supporters remained active in the constituencies.

Yet the progress of the left on the policy front was reflected in the preparation of the manifesto for the 1983 general election.[27] By tradition Conference resolutions set the terms for negotiations between the NEC and the PLP; under Clause 5 of the party constitution these two bodies jointly approve the manifesto. Under Mr Wilson and Mr Callaghan this meeting had been used by the parliamentary leadership to veto proposals which it found objectionable. In 1973, for example, Mr Wilson had boldly rejected the NEC's proposals to nationalize the twenty-five largest companies in Britain and threatened to 'veto' them in the Clause 5 meeting and in 1979 Mr Callaghan had refused to include in the manifesto a number of policies which were popular with the left and supported by the annual Conference.

In 1983 a campaign document entitled *The New Hope for Britain* was agreed by the NEC and the Shadow Cabinet. It contained many policies approved by Conference, such as withdrawal from the EEC, commitment to unilateral nuclear disarmament within the lifetime of the next Parliament, plans for a massive extension of public ownership, and increased public spending; and it avoided a commitment to an incomes policy. All of these were known to be anathema to many in the parliamentary leadership. On the eve of the general election Mr Michael Foot suggested to the Shadow Cabinet that the document be largely accepted as the manifesto. There was little discussion of the draft document in the Clause 5 meeting and it was forced through in an hour or so, virtually undiscussed. It was the shortest Clause 5 meeting ever and showed no evidence of parliamentary domination. The memory of Mr Callaghan's behaviour in 1979—and the rows it led to—served as a warning symbol. That legacy and the shift in the balance of power greatly weakened the hands of the parliamentary leaders.

The Party Split

The 1979 Parliament was significant for a split in the Labour Party and the emergence of a substantial 'third force' in British politics. Since the exit of the Irish Nationalists from Parliament in 1918 and the decline of the Liberals after 1929 Britain effectively had had a two-party system. Although the Liberals made a gradual recovery and reached 18 per cent of the vote in February 1974 the non-proportional electoral system was a barrier. There had been episodic talk since the 1960s of a realignment on the centre-left, involving a fusion of Liberals and Labour right-wingers but this was usually dismissed as a pipe dream by Labour leaders. For example, Roy Jenkins in 1973 had dismissed talk of such a centre party: 'I do not believe that such a grouping would have any coherent philosophical base . . . A party based on such a rag-bag could stand for nothing positive.'[28]

But during 1979–80 three forces combined to assist the creation of the Social Democratic party.[29] The first was the availability of Roy Jenkins, whose term of office as President of the European Commission was due to end in January 1981, so that he was looking for a means of re-entry to British politics. In 1976 he had left a post in the Labour Cabinet to become the Commission's President, but gradually became disenchanted with the Labour party's lurch to the left.

The second element in the process of mould breaking was the gradual disillusion of substantial figures in the Labour party with what was happening to the party. In the end it was the 'Gang of Three', Shirley Williams (out of Parliament), Bill Rodgers, and David Owen who opened up the path to a breakaway party. These were all members of the previous

Labour administration who could reasonably have expected to hold high office in a future Labour government. The attempts of the left to promote policies which ran counter to those of the previous administration were at first met with warnings from the 'Gang of Three' about their collective determination to fight back and about the electoral damage that the left was doing to the party. But it was the success of the left in forcing through the constitutional changes that was the final straw for many of the right. It was then that the three realized that they might want to leave and form a new political party. On 7 June 1980 they issued a statement arguing that Britain must stay in the EEC and, in the final paragraph, hinted that they might have to leave the Labour camp.

On 25 January, the day after the Labour party's Wembley Conference, they launched the 'Limehouse Declaration' and were supported by nine other Labour MPs. The declaration was a statement of social democratic values and policies endorsed by a hundred names, mostly middle-aged, right-wing Labour figures who had been prominent a decade earlier. Many of the ideas were highly consensual, redolent of the 1960s and of Harold Wilson's first government. In March the new Social Democratic Party was formally launched, supported now by fourteen ex-Labour MPs and one Conservative, Christopher Brocklebank-Fowler. Within twelve months the number of MPs grew to twenty-nine, thanks to further defections from Labour and two by-elections.

Opinion polls and by-elections showed a remarkable degree of support for the new party. The immediate success of the party has largely to be explained by its attractive personalities and the sympathetic mass media coverage accorded to them, as well as by the continuing inability of the Conservative party to produce economic success or of the Labour party to stop its internal disputes. Each new defection of a Labour MP made headlines.

During 1981 and 1982 the Alliance of Social Democrats and Liberals offered the most formidable threat to the two-party system in over fifty years. But the major problem for the Alliance was the weak partisanship of SDP and Liberal voters. Its fortunes would depend upon the other parties. It had not mobilized a clear social or geographical constituency nor occupied a distinct issue space. Once the element of novelty had disappeared, the Alliance found it difficult to secure the attention of the mass media. Within the first twenty months of the SDP's formation, support in opinion polls for the new party and the Alliance fluctuated between 50 and 20 per cent, a telling indicator of the 'softness' of the party's base. In early 1982 a downturn in the Alliance's fortunes was apparent even before the Falklands episode.

Conclusion

In the 1983 election Labour's electoral decline continued and even accelerated. Its share of the vote, 27.6 per cent, was a fall of 16.2 per cent from that in its clear defeat in 1959 and of 11.9 per cent from its last victory in October 1974. In terms of votes cast per candidate it was the worst result since 1900, when the party was born. It was now a party of a section of the working class; only 38 per cent of the manual workers and 39 per cent of trade unionists voted for it (compared to 32 per cent for the Conservatives). It had the support of workers who lived in Scotland and the north of England, worked in the public sector, and lived in council homes. But affluent workers, who were car-owners, buying their own homes, and employed in the private sector, clearly preferred the Conservatives and had economic incentives for doing so. The Conservatives enjoyed an eight-point lead in the skilled working class (40 to 32 per cent) according to Gallup. In its share of the vote in 1983 it trailed behind socialist parties in France, Austria, Denmark, Norway, Sweden, Spain, Portugal, Greece, West Germany, and Belgium.

It is interesting to speculate why the party appeared to be so out of touch with its supporters. Why did the Downsian model, which assumes that party leaders adopt policies which most voters want, break down? One explanation must take account of the revival of the idea of inner party democracy in the Labour party in the 1970s. The reforms after 1980 decentralized power to the constituencies and the trade unions. Powers were dispersed to activists who were not elected by the electorate and against whom no electoral sanction could be wielded. Downs assumes that power is concentrated in the hands of the elected politicians, who therefore have an incentive to be responsive to voters. In the case of the Labour party the power was dispersed to activists who, on many issues, were out of sympathy with the views of the PLP. The radicalization of the party found little echo in the country. The 'gap' between the party's policies, increasingly shaped by Conference, and the views of Labour voters widened in 1979 and 1983. Invariably, the party policies were to the left of what voters wanted. On six of eight important issues in the 1979 general election—policies for unemployment, industrial relations, incomes policy, public ownership, social services, race relations, taxation, and the European Community—the Conservatives were more representative than Labour not only of the electorate but also of the working class.[30] The same was true in 1983.[31] If anything, Labour became more of a vanguard party, rejecting the assumption that the electorate itself knew what was good for it.

A number of factors were at work to produce the shift in the Labour party. The charge of the leadership's 'betrayal' of socialism has been a persistent theme of the left wing in the history of the party. After 1979, however, it was more sustained, more widely articulated throughout the movement, and linked to demands for constitutional changes. Changes in the trade unions—particularly the shift to the left in many executives—made the unions less reliable allies for a right-wing parliamentary leadership at Conferences. The fact that Labour governments since 1964 had also trespassed on the unions' sacred ground of free collective bargaining only added to the tensions. After 1979 the party not only moved further to the left but power also shifted in the direction of Conference. The view that revisionist and consensus policies had been tried and failed gained ground.

These changes coincided with the party's electoral decline. It was not easy to argue which was cause and which was effect, as the party had been in decline for some years. A front-bench spokesman, Peter Shore, claimed that the electoral rout in 1983 meant that Labour had 'lost the 1980s'. Was Labour to change its policies, particularly on defence, council housing, and public ownership and adjust to an affluent working and more right-wing electorate? Or should it, like Tony Benn, take comfort that 'for the first time since 1945 a political party with an openly socialist policy has received the support of over $8\frac{1}{2}$ million people'.[32] The left could argue that the old consensus had broken down, that the revisionists had either departed (some to the SDP) or were no longer so influential in the party. But the main beneficiary was the Thatcher wing of the Conservative party.

In studying the decline of the Labour party this chapter has dealt with the interaction of four different factors. First, there has been the changing role of the trade unions. Second, there was a negative reaction to the records of the Wilson and Callaghan administrations. Third, there was the decline in the influence of revisionist ideas and the growth of a more radical left wing. Fourth, there were the constitutional changes in 1981 which gave the extra-parliamentary party and the left wing a greater say in the party. They also precipitated the exodus of a number of right-wingers to form the Social Democratic party. From 1970 the revisionist wing had been in retreat, although a Labour Cabinet was still effectively dominated by right-wing figures. But the Winter of Discontent, the election rout in 1979, and then the steady decline in the authority of the parliamentary wing and the exit of some right-wing leaders to the SDP shifted the balance in the party. Defenders of the mixed economy or of the records of previous Labour governments were isolated voices. Labour had become more radical just as had the Conservatives under Mrs Thatcher.

Notes

1. *Politicians and the Slump* (London, Macmillan, 1967).
2. *The Future of Socialism* (London, Cape, 1956).
3. Ibid., p. 517.
4. Ibid., pp. 28–9.
5. M. Abrams and R. Rose, *Must Labour Lose?* (London, Penguin, 1960).
6. (London, Macmillan, 1969), p. 175.
7. *The Labour Party Conference* (Manchester University Press, 1980).
8. *British Social Democracy* (London, Croom Helm, 1980), p. 282.
9. See D. Kavanagh, 'Power in British Parties, Iron Law or Special Pleading?', *West European Politics* (1985) and R. McKenzie, *British Political Parties* (London, Heinemann, 1955).

10. M. Harrison, *Trade Unions and the Labour Party* (London, Allen & Unwin, 1960).
11. *The Labour Party Conference*, p. 326.
12. An influential statement of this thinking was by S. Holland, *The Challenge* (London, Quartet, 1975).
13. For a good account of this episode, see P. Whitehead, *The Writing on the Wall* (London, Michael Joseph, 1985), chap. 9.
14. D. Coates, *Labour in Power* (London, Longman, 1981).
15. R. Miliband, *Parliamentary Socialism* (London, Allen & Unwin, 1961), p. 281.
16. *Labour in Power*.
17. D. Butler and D. Stokes, *Political Change in Britain* (London, Macmillan, 2nd edn. 1974), pp. 296–302.
18. J. Alt, *The Politics of Economic Decline* (Cambridge University Press, 1978), p. 281.
19. A. Crosland, *Socialism Now* (London, Cape, 1975).
20. A. Arblaster, 'Anthony Crosland: Labour's Last Revisionist', *Political Quarterly* (1977), J. Mackintosh, 'Has Social Democracy Failed?', *Political Quarterly* (1978).
21. I. Crewe, 'Labour and the Electorate', in D. Kavanagh (ed.), *The Politics of the Labour Party* (London, Allen & Unwin, 1982).
22. J. Gyford, *The Politics of Local Socialism* (London, Allen & Unwin, 1985).
23. See P. Whiteley, *The Labour Party in Crisis* (London, Methuen, 1983).
24. A. Mitchell, *Four Years in the Death of the Labour Party* (London, Methuen, 1983), p. 32.
25. For example, R. McKenzie, *British Political Parties*.
26. D. Marquand, 'Inquest on a Movement', *Encounter* (1979), p. 14.
27. This section draws on D. Butler and D. Kavanagh, *The British General Election of 1983* (London, Macmillan, 1984).
28. *What Matters Now* (London, Fontana, 1972).
29. I. Bradley, *Breaking the Mould* (Oxford, Martin Robertson, 1982).
30. I. Crewe, 'Labour and the Electorate', pp. 28–38.
31. I. Crewe, 'How to Win a Landslide without Trying', in A. Ranney (ed.), *Britain at the Polls 1983* (Duke University Press, 1985).
32. A. Benn, 'Spirit of Labour Reborn', *The Guardian*, 20 July 1982.

7

A MORE RESOLUTE CONSERVATISM

THE Conservative party was in government for more than half the years between 1945 and 1974 and consolidated many parts of the post-war consensus. There had, however, always been Conservatives who objected to the drift of the party's post-war policies which they regarded as being too collectivist or 'socialist'. But they were usually a beleaguered minority among the party's leadership. The significance of Mrs Thatcher is that she shared many of the values of this traditional minority. This chapter explores the tensions in the party which came to a head under Mr Heath and how Mrs Thatcher managed to lead the party in a different direction from her predecessors.

By comparison with the vast literature on the Labour party, the Conservative party has been much less studied.[1] One reason for this is that the Conservative party has so often been in government, and British government is highly secretive. By tradition the party leaders have also been reluctant to expose its innermost workings to the outside world. Only in 1965 did MPs and Cabinet Ministers have a formal right to participate in the election of the party leader. There is no equivalent of the indiscreet 'tell all' diaries of former Labour ministers like Dalton, Crossman, and Mrs Castle.[2]

Yet the party has probably been more influential in shaping British political ideas and political institutions than has the Labour party.[3] Party spokesmen regularly claim that they are the national party. In his book *The Case for Conservatism* Quintin Hogg said, 'Being Conservative is only another way of being British.'[4] It has also been the most electorally successful right-wing party in the twentieth century. In the hundred years since 1885 it has been in office for all but twenty-eight years. On only three occasions since then—1906, 1945, and 1966—has there been a British Parliament in which the party of the left had a clear working majority sufficient to last for a full Parliament. In contrast to various right-wing political parties in Western Europe it was not tinged by association with Fascism before the war, nor until recently with right-wing extremism; it has not had to rely upon a large agricultural or peasant base for electoral support; and it has not drawn on the support of a powerful Roman Catholic Church, as the political right has in France and Italy. The Conservative party has also enjoyed a number of other advantages. The political opposition has often been split, as for example the Liberals were in 1885 and again in 1916, and Labour was in 1931 and again in 1981. As the largest party at a time of such fragmentation it has gained from the working of the first past the post electoral system. As a party of the right it has also profited from the sheer continuity of the British regime. When regimes elsewhere in Europe were overthrown by revolution or defeat in wars, so the ruling élites were frequently discredited. In 1944 in France Petain dragged the traditional élite down with him, while in 1940 Churchill managed to save the British Establishment. The Conservative party has managed from the late nineteenth century to provide a home for commercial, business, and landed interests. It has also come to terms with the growing power of the state and the emergence of a mass electorate. It has been lucky, opportunistic, but also skilful.

Conservatives have shown a remarkable concern with winning and holding office and have given a high priority to strong and authoritative leadership. The Labour party's arrangements for regular election of the party leader and election of the Shadow Cabinet by MPs and the expectation among the grass roots that the party leadership is accountable to the annual Labour Conference was a negative example for Conservatives. The Conservative Conference still makes no pretence to being a decision-making body for the party. The pursuit of office and support for authoritative leadership have facilitated party unity and the leadership's ability to be adaptable on policy questions.

In a famous column in 1883 *The Times* claimed that Disraeli had discerned in the English working class 'the conservative working man as a sculptor perceives the angel imprisoned in a block of marble'. A number of social, psychological, and sociological studies have explored the nature of this working-class support for the party. Without these 'deviants', usually amounting to about a third of the electorate, Britain would not have had a balanced two-party system. Some students have emphasized the importance of social deference as an explanation of working-class Conservatism; this refers to the phenomenon of workers who regard the possession of high status education and social background and other signs of superior breeding as proof of political ability.[5] Such voters are inclined to regard the upper class as 'natural' political rulers, or claim that 'leadership is in the blood' and vote Conservative. Other studies have emphasized that material affluence or the possession of so-called middle-class attributes such as home ownership or car ownership is an incentive for workers to vote Conservative. Another theory links Conservative support with old age, not on the grounds that voters become more politically Conservative as they grow older but that the party has for long had a strong following among the old because it was established before Labour became an effective contender for electoral power. More instrumental reasons for workers supporting the party are that they prefer its leadership, governmental record, and policies.

It is often claimed that the party is linked to a ruling class in Britain. Such institutions, according to some commentators, as the Church of England, the BBC, the bulk of the national press, the ancient universities, the professions, the City, and private business have often been associated with the Conservative party.[6] There are numerous studies showing the similarity between the upper-class, socio-economic backgrounds of Conservative leaders and of many other élites; indeed, it has sometimes been claimed that the party serves these particular interests. Members of the major institutions of the British state—the judiciary, the national press, business, finance, and the civil service—are frequently alleged to be more in sympathy with the party. The fact that the party has often been in government, and therefore works with people in these institutions, has reinforced the suspicion. The connection of social, political, and economic status may have reached a high point in the Conservative Cabinets of the 1950s. Churchill, for example, had five peers in his 1951 Cabinet of sixteen, and in

1962 a third of Macmillan's government could trace a relationship to the Duke of Devonshire.

There is a remarkable consistency of view among writers on the character of British Conservatism. The party lacks a doctrine or settled principles—indeed most Conservatives derive consolation from this. There is a long list of policies and principles, protection in the nineteenth century, union with Ireland, British Empire, powers of the House of Lords, free enterprise (that is, no state ownership) and free collective bargaining, and so on, which the party has been pledged to uphold and has subsequently abandoned. Freedom from ideology, attachment to the importance of retaining office, and responsiveness to 'circumstances' have made it the adaptable party. Conservatives draw freely on the writings and behaviour of Burke, Hume, Coleridge, Disraeli, and Salisbury. Conservatism is said to have its own logic and to be closely related to political practice, to be an 'attitude', 'a habit of mind', or even 'a way of life'. It includes respect for the traditions of society, rule of law, private property, and individual liberty. It has also, at least until the mid-1960s and mid-1970s, been fairly sympathetic to the status quo. The party has been prepared to introduce reforms but only gradually and cautiously.

Conservative Tensions

It is usual to discern two main strands operating in British Conservatism, the liberal and the collectivist, sometimes called 'neo-liberal' and Tory tendencies respectively. For much of its history the party has combined these two strands; at times the first strand has been uppermost, at other times the second. The relative influence of the strands has depended, in part, on the values of the political leadership at any time (for example, Mrs Thatcher's views differ in many respects from Mr Macmillan's) and on political circumstances. A remarkable example of the same administration moving from one strand to another is seen in the shift in policy of the Heath government between 1970 and 1974. The administration started out with a neo-liberal set of policies for managing the economy but by 1972 had made spectacular

U-turns, introducing far-reaching statutory controls on prices, incomes, and dividends and assuming extensive powers for intervention in industry. More usually, as Greenleaf notes, the strands are fused in the practice of a Conservative government.[7] A recent study of the different clusters of Conservative ideas concludes that the diversity is such that it is implausible for any group to claim that it is following *true* Conservative principles.[8]

The neo-liberal outlook emphasizes the importance of the individual and the limited role of government and clearly derives from classic liberalism (see Chapter 4, above). Although government intervention in society and the economy can be justified on grounds of, for example, protecting social order and removing demonstrable social evils, as a rule the case has to be made for government intervention. More recently, as we saw in Chapter 3, a new right has borrowed the neo-liberal language and become more influential in the party. It talks more confidently about the virtues of individualism, free markets, and competition, as opposed to those of tradition, social cohesion, and compassion. In the mid-1960s the main spokesman for this point of view was Enoch Powell, who called for major cuts in direct taxation, a rolling back of the welfare state, and denationalization of the nationalized industries (see above pp. 69–70). The collectivists have generally stressed the importance of community and made a positive case for the use of public power to promote the general interest, which they see as emerging from purposive state action rather than the free interaction of individuals. Samuel Beer has noted how this aspect of British Conservatism joined with socialism in the post-1918 period to accept a more activist role for the government as a means of promoting the shared goals of the community.[9] But, in making the case for government intervention, Conservatives have differed from socialists, justifying it by reference to the authority of government *per se* rather than an election manifesto, and to the interest of the nation rather than of 'the people' or a social class. On the whole, most accounts of British Conservatism in the twentieth century have seen the collectivists as being dominant, partly for reasons of government and partly for reasons of electoral expediency.

The two different strands do not easily parallel differences between the political left and the right in the party. There were, for example, a number of Conservatives who were sceptical about the wartime Beveridge proposals. 'Die-hards' like Sir Herbert Williams and Sir Waldron Smithers opposed the Beveridge document as 'a very bad report' and were also critical of the Industrial Charter of 1947 which signified the party's acceptance of the new mood.[10] In 1945 many Conservatives were still reluctant to accept the financial burdens entailed by the welfare reforms. Beer's view was that the party had not yet made a clear commitment to the Managed Economy or Welfare State.[11] In the 1950s the 'die-hards' became more concerned about the retreat from Empire and there was a small Suez group opposed to British withdrawal from that zone. But these MPs did not overlap with free-marketeers like Enoch Powell. Indeed, a study of back-bench opinion in the party in the 1950s found little evidence of organized factions; rather, there were tendencies in which different groups of MPs joined together on different issues.[12] Free-marketeers became increasingly impatient with the policies of the Macmillan and eventually the Heath governments. Robert Behrens's study of the tensions between 'Diehard' and 'Ditcher' groups in the party claims that by 1975; 'The Diehards viewed post-war Conservatism as merely an alternative form of socialism. The problem was that Conservatives had deviated from the eternal principle of limited government, sound money and moral rectitude.'[13]

The One Nation form of Conservatism was represented by a number of MPs who were part of the 1950 election intake. They included men like Ted Heath, Reginald Maudling, Iain Macleod, Robert Carr, and Enoch Powell, all of whom later achieved full Cabinet positions. They saw themselves as representing an established Tory tradition going back to Disraeli and heeding his warning that the Tory party is a 'national party or it is nothing'. For an inspiration they could point to the social reforms passed by Disraeli's administration between 1874 and 1880, Baldwin's acceptance of the Labour party in the 1920s, and the social reforms and the managed economy of the 1930s. R. A. Butler and Harold Macmillan speedily came to terms with the post-war policies of the

Labour government. Butler in a speech in 1946 called for 'an acceptance of redistributive taxation . . . and repudiation of *laissez faire* economics in favour of a system in which the state acted as a trustee for the interests of the community'.[14] In 1947 the policy document The Industrial Charter accepted full employment, public ownership, and proposed a new charter of rights for workers. A remark attributed to R. A. Butler in 1947 was 'if they (the people) want that sort of life (welfare and high public spending) they can have it, but under our auspices'.[15] The Maxwell-Fyfe reforms of the party, which prohibited large contributions by candidates to the constituency association, opened the way for young men and women without wealth to become Conservative candidates.

Conservative Governance

The 1950s were a golden age for the Conservative party, in electoral terms at least. The party's majority of seventeen seats in the 1951 election was increased in 1955 and again in 1959. The watchwords of the new Conservatism were conciliation, moderation, and consensus. Churchill appointed Walter Monckton as Minister of Labour, with a brief to maintain good relations with the trade unions. (He was described by a colleague as 'that old oil can'.) There was a relaxation of controls over industry and an end of austerity policies, although these policies were already in train before Labour left office in 1951. Labour's programme of nationalization was accepted, except in the cases of iron and steel and road haulage. The Conservative government was also able to maintain spending on welfare, partly out of economic growth and partly out of a reduced defence commitment, so that not until 1957 did public spending increase as a proportion of GDP.

After 1959 Conservative ministers, particularly Macmillan, became more concerned over the stop-go economic policy and the evidence that Britain was falling economically behind her Western European neighbours. As part of the great reappraisal of policy in 1961 Britain applied to enter the European Community and moved towards indicative economic planning and incomes policy as stimuli to greater growth. Macmillan's government was interventionist towards industry, retained many controls, protected monopolies, and restricted competition. Not until 1963, for example, was a determined minister, Mr Heath, prepared to abolish resale price maintenance. Nigel Harris points out that 'the economics of neo-liberalism diverged increasingly from the politics of government'.[16] In Cabinet discussions in 1961 about proposals to introduce economic planning and establish the National Economic Development Council, Macmillan noted in his diary: '. . . a rather interesting and quite deep divergence of view between ministers, really corresponding to whether they had old Whig, Liberal laissez-faire tradition, or Tory opinions, paternalist and not afraid of a little *dirigisme*'.[17] He was firmly a member of the latter group.

The formation of the Monday Club in 1961, in reaction to Macmillan's 'Wind of Change' speech in Africa, illustrated the complexity of the policy differences in the Conservative party. The group felt that the post-war party under Butler and Macmillan had compromised too much with socialism. It urged the party to reassert the importance of individualism, take a strong line on law and order and immigration, and repatriate all non-European immigrants. But, unlike many free-market advocates, it was nationalist and resented Britain's retreat from Empire. What was needed was 'genuine Conservatism', which would overturn the post-war consensus. According to its newspaper, *Monday World*: 'The "opinion formers" and the politicians can no longer ignore the increasing demands of the people. They are calling for stricter control of immigration to protect standards of education and housing; they are seeking sterner penalties for criminals in the face of the phenomenal increase in crime; they are demanding discipline in the universities.'[18]

Right Turn

An explanation for the dominance of this Tory strand in the post-war period may be that the party was in office from 1951 until 1964. Certainly a different emphasis was apparent after 1964, when the party was in opposition, particularly after Mr Heath became leader in 1965. The Conservative party in government appeared to have already run out of ideas by 1964 and there was a feeling among some leading Conservatives that the moderate political establishment of both major parties had missed too many opportunities over the previous two decades. In the party manifesto for the 1966 election, *Putting Britain Right Ahead*, a number of new policies were prepared, many of which foreshadowed Mrs Thatcher. There were proposals to reduce direct taxation, provide greater incentives for managers, break up monopolies and increase competition, enter the Common Market, and move to a selective provision of welfare benefits. Another theme of the future Heath government, reform of trade unions, soon followed. After the 1979 general election *The Economist* (26 May 1979) published a mock Queen's Speech in which passages from Mr Heath's first (1970) Queen's Speech, were mixed with those of Mrs Thatcher's government in 1979. It was extremely difficult, given the similarity in statements, to guess which was which.

Some Conservatives became increasingly dissatisfied with the disorder in the labour market. As early as 1958 a group of Conservative lawyers, in *A Giant's Strength*, had proposed reforms to outlaw strikes which were political and breached a trade union's rules, require a 'cooling off' period before strikes began, and curtail the operation of the closed shop. These critics came into their own when the party was in opposition. The party proposed to remove from trade unions the immunities which they enjoyed under common law and to encourage them to control their members. The leadership hoped that the avoidance of incomes policies would be a sweetener for the unions and encourage them to concentrate on wage bargaining. Although the party was hoping to depoliticize industrial relations, Colin Crouch suggests that the attempt to use law in industrial relations was a major breach in the post-war consensus.[19]

The central objective of the rethinking of policy was to promote greater economic efficiency and productivity. It was also coloured by a managerialist orientation. In 1965 Tim Raison, a young Conservative MP, wrote in a Conservative Political Centre pamphlet, *Conflict and Conservatism*, that 'we should regard the Conservative Party in the economic context as the management . . .' Compared to the Macmillan–Butler era there was also a retreat from reliance on incomes policy and economic planning. Although many of the policies resembled those of the Conservative party under Mrs Thatcher there were also important differences. Mrs Thatcher justified many of the policies on the grounds that they promoted economic freedom, as well as prosperity; for Mr Heath they were instrumental in assisting economic growth; they were good if they worked. Mrs Thatcher has a more deep-rooted dislike of an economically interventionist state—though not, of course, of the state as such (Chapter 9). Mr Heath was a technocrat who thought that government measures could actually encourage individual enterprise to flourish. Mrs Thatcher objected to the level of direct taxation, whereas Mr Heath was more concerned to change the balance between direct and indirect taxes. Mrs Thatcher has never been much interested in reforming political institutions, whereas Mr Heath was a great believer in reforming institutions and structures as a way of improving performance (witness his introduction in 1970 of the Central Policy Review Staff and the creation of super-departments).

There is still a good deal of controversy, not to say bitterness, about the record of the Heath government. In 1970 Mr Heath's set of policies appeared to mark a clear break with Conservative traditions. The reduction of state intervention in the economy, avoidance of a formal prices and incomes policy, cuts in public expenditure and direct taxation, greater selectivity in welfare, and the penalties proposed for breaches of the proposed industrial relations law were all seen as breaches with the post-war approach. Joining the European Community also marked a break with the Atlantic–Commonwealth outlook that had previously been so strong in the leaderships of both parties. At the time the economic and social policies were regarded by a number of commentators as a neo-liberal challenge to the post-war collectivist consensus.

Within two years, however, the Heath government had to perform two spectacular U-turns, adopting an elaborate statutory prices and incomes policy, and intervening to rescue firms facing bankruptcy. The 1970 manifesto bluntly stated, 'We utterly reject the philosophy of compulsory wage control', and leaders promised that they would not bail out lame duck companies. Rising inflation and unemployment, however, posed a problem for the government's declared economic strategy. The government nationalized the Rolls–Royce aerospace engine company to save the firm from bankruptcy in 1971, and the Industry Act (1972) gave it sweeping powers to intervene in industry. The coal miners challenged the third stage of the incomes policy in the winter of 1973 and by working to rule and then striking caused widespread industrial disruption. Under pressure, Mr Heath felt compelled to call a general election in February of 1974 to defend his government's statutory incomes policy. He lost. Another general election was held in October 1974 and the Conservatives this time fought on a national unity and coalition platform. The election resulted in an even heavier defeat for the party.

Views of the 1970–4 record have played an important part in colouring assessments of Mrs Thatcher.[20] It was probably during the Heath administration that the Conservatives lost, temporarily at least, their reputation as the party of competent government. The bitter industrial conflicts, the highest number of days lost in strikes (in 1970, 1971, and 1972) since the 1926 general strike, gathering inflation, three-day week, and crisis general election because of the miner's strike all suggested a government under siege. By 1974 the Emergency Powers Act had been invoked a total of twelve times since it was introduced in 1920; five of these had been by the Heath government in connection with strikes. By so eagerly seeking the co-operation of the TUC for an incomes policy the leadership was in danger, according to Tory critics, of compromising the authority of government.

Supporters of Mr Heath allege that Mrs Thatcher has been selective in her memory of the 1970–4 government and has attempted to rewrite history. Pointing to the similarity between Mr Heath's initial policy objectives and those of Mrs Thatcher, supporters of the latter have diagnosed the major failure of the Heath government as one of lack of political will. There was, however, a different inspiration behind the programmes of the two leaders. Mr Heath regarded the policies as *means* to solving problems and promoting economic growth; if they failed, then something else could be tried. The Heath government still accepted Keynesianism and the goal of full employment. The so-called 'Barber boom', involving reflation by expansionist monetary and fiscal policy, was an attempt to make the old system work. After he lost the party leadership Mr Heath in his turn became bitter at the way some former Cabinet colleagues, notably Mrs Thatcher and Sir Keith Joseph, were prepared to disown their record in his government. But some Conservatives, particularly outside the Cabinet, had been doubtful of the statutory incomes policy from the start. They did not consider it was possible in a free society to legislate for income differentials, prices, and dividends.[21]

Between 1931 and 1964 the Conservatives had lost office only twice (1945 and 1964). By 1974, however, the sense that the party was the natural party of government had been severely shaken, not least among Conservatives themselves. Labour emerged successful from four out of the five general elections held between 1964 and 1974 and had been in office for all but three and a half of the fourteen and a half years from 1964 to 1979. In addition, the political agenda placed the Conservative party at a disadvantage. For a number of years governments of both parties had pursued incomes policies and sought the goodwill of the unions in the battle against inflation. The trend to interventionist and collectivist policies and the enhanced bargaining power of the unions were more in tune with Labour than Conservative interests.

After the second election defeat in 1974 the Conservatives faced a number of critical decisions. Under Mr Heath the party had lost three of the previous four general elections. Moreover, it seemed to have lost its way on managing the economy, particularly in mounting a credible anti-inflation policy. Mr Heath was firmly associated in the public mind with a statutory prices and incomes policy. The choice seemed to be between a formal policy of incomes restraint, which required the co-operation of the unions and granting them an important say on government policies, or monetarism, which would permit free collective bargaining, but also produce high unemployment, thus weakening the bargaining power of the unions. The latter policy was generally thought to be 'unworkable' both because of the political unpopularity the government would incur and also because the Labour party had already won the consent of the trade unions with its social contract.

Mr Heath rebuffed early pressures, including some from his friends, that he should either offer himself for re-election as leader or resign. Under the 1965 rules of the party, which provided for an election but not for re-election of a leader, there was no way of forcing the issue. In the four months after the October general election there was something of a vacuum in the Conservative leadership. The issue was broached in the executive of the 1922 Committee which met on 14 October at the home of its chairman, Edward Du Cann. The Committee unanimously agreed that there should be a leadership election in the new parliamentary session and Mr Du Cann was asked to convey this message to Mr Heath. Unfortunately the relations between the two men had been cool since Mr Du Cann's time as party chairman under Mr Heath in 1966–7 and tension was increased by media speculation that Mr Du Cann himself was a contender for the leadership.

Mr Heath refused to discuss the leadership issue with the executive of the 1922 Committee until the Committee had an election for new officers. Yet he misread the mood of the party. Under further pressure a committee was established under Lord Home to consider changes to the rules established in 1965 for the election of a Conservative leader. Three main proposals for change were accepted. First, there should be an annual election of the leader; second, a candidate, in order to be elected on the first ballot, required not only an overall majority but also a lead over the runner-up equal to 15 per cent of those eligible to vote; and third, that only MPs would have votes although the views of the Conservative peers and of the constituency associations should be conveyed to them.

Thatcher's Election

Mrs Thatcher was something of a political unknown in early 1975. She had been in the Cabinet as Secretary of State for Education but this was not a senior post and she was not close to Mr Heath. She had played a prominent part in the October 1974 election campaign, with a pledge to reduce the mortgage rate for home-owners to 9½ per cent or less (originally forced on her by Shadow colleagues), and impressed her colleagues when she led the parliamentary opposition to the Finance Bill in early 1975. After the October election defeat she was a supporter of the leadership claims of Sir Keith Joseph. Sir Keith, who had also served as a Cabinet minister under Mr Heath, was rethinking Conservative principles during 1974, and he soon emerged as the most prominent free-market spokesman for those who disowned Mr Heath's economic policies (see Chapter 4). It was only after Sir Keith declined to stand against Mr Heath that Mrs Thatcher considered running herself. Other former Cabinet ministers, including the Deputy Leader and obvious heir-apparent William Whitelaw, felt bound both by ties of personal loyalty to Mr Heath and of collective responsibility for the record of the last government, and refused to stand.

Margaret Thatcher was a surprising person to mount a challenge to Mr Heath. She came from a small town, Grantham, in the East Midlands (once voted the most boring town in Britain). Her parents had run a small grocer's shop and her father had been a local councillor, although by becoming Mayor of Grantham he was a substantial local figure. By dint of hard work, earnestness (she had little time for social frills), and brains, she won a place at Oxford in 1943, qualifying first as a research chemist and then, after a spell in industry, as a tax lawyer. She was obviously a very determined woman. Like politics, both fields were and still are male-dominated and rather unusual careers for a woman to pursue. This may have confirmed the qualities that some identify as hardness—a belief that ability and hard work can enable a person to get to the top. In politics she never seems to have traded on being a woman. She married a wealthy businessman, Denis Thatcher, and his money provided her with the opportunity to pursue a career as a professional politician. She entered parliament in 1959 at the age of 34 and, with the Conservatives in opposition (1964–70), was appointed as Shadow Cabinet spokesman for education in 1969. It was ironical that in this post she replaced the epitome of Tory progressivism, Sir Edward Boyle. On becoming Prime Minister in 1970, Mr Heath appointed her to the Cabinet post of Secretary of State for Education and Science, a post which she held until the fall of the government.

Mrs Thatcher had therefore been an MP for 'only' fifteen years (1959 to December 1974) when she challenged Mr Heath and her previous ministerial experience at Cabinet level had been confined to a relatively minor department. British Prime Ministers have usually ascended a ministerial hierarchy and proved their political and administrative ability to parliamentary colleagues over many years. In comparison with all other post-war Conservative leaders, Mrs Thatcher was a 'newcomer'.

In standing for election she drew support inevitably from diverse quarters, from those who felt slighted by Mr Heath, from those who were critical of the U-turns and policy reversals made by his government, from a number of right-wingers, from disillusioned Keynesians, and from those who simply felt it was time for a change of a leader if the party was to win the next election. Her backers, led by Airey Neave, deliberately 'talked down' her chances so as to maximize the full anti-Heath vote. Attempts were subsequently made to invest Mrs Thatcher's election as leader with ideological significance. It is true that many of her free-market instincts were already present. But they were articulated more systematically after she became leader. King fairly writes that 'she was not elected as a Thatcherite'.[22] The decisive consideration was that she was the only serious candidate to oppose Mr Heath and it was necessary for MPs to vote for her, though not necessarily for her ideas, if they wanted a change. She was something of an accidental leader, and it is a rewriting of history to see her election as a victory for monetarism. Apart from a few MPs like John Biffen, Enoch Powell, Sir Keith Joseph, and Jock Bruce-Gardyne, and a few academics and economic commentators, there was little talk at the time about monetarism.

In a new election which was held on 4 February, she defeated Mr Heath by 130 votes to 119. Mr Heath had had the support of most of the senior figures in the party and of the bulk of the peers and, apparently, the sympathy of most of the constituency associations. But he had not cultivated the backbenchers, a lesson which was not to be lost on Mrs Thatcher. He withdrew from the election process but a second ballot was necessary because Mrs Thatcher had not reached the target figure of 139 votes (that is, an overall majority of votes and a lead of at least 15 per cent over the runner-up). The way was clear for a larger field and Mr Whitelaw, Mr Prior, and Sir Geoffrey Howe entered. But the second ballot, held on 11 February, proved decisive and Mrs Thatcher easily defeated Mr Whitelaw to become the first woman leader of any British political party (see table 7.1).

TABLE 7.1 *Conservative Leadership Election 1975*

First Ballot (4 Feb.)		Second Ballot (11 Feb.)	
Mrs Thatcher	130	Mrs Thatcher	146
Edward Heath	119	William Whitelaw	79
Sir Hugh Fraser	16	James Prior	19
Votes not cast	11	Sir Geoffrey Howe	19
		John Peyton	11
		Votes not cast	2

Although front bench Conservatives remained predominantly upper-middle-class products of public schools and Oxbridge, Mr Heath and Mrs Thatcher represented a more meritocratic strain in the Conservative party. Both came from comparatively modest family backgrounds, won scholarships to local grammar schools and Oxford, and were self-made professional politicians. They belonged to the 'tough-minded' rather than the 'tender-minded' wing of the party. Both were impatient with the status quo and favoured more radical free-market solutions to the problems of the economy. But there were some important differences. Although he was the first leader chosen by MPs in the contested election, Mr Heath had long been a member of the party's leadership strata, ascended the ministerial hierarchy, and might well have 'emerged' under the old system of informal consultation. He also represented continuity with many of the 'One Nation' principles associated with Sir Anthony Eden, Macmillan, and Butler. Mrs Thatcher, however, had overthrown an established leader, had not previously served in any of the major offices of state, was a product of a back-bench rebellion against the leadership, and lacked weighty front-bench support on the first ballot. (Similarly, in 1922 Bonar Law's election to the party leadership, after Lloyd George's resignation, was the product of a back-bench reaction against the existing party leadership. He was shortly followed by another 'outsider', Stanley Baldwin, who was helped by the fact that the most favoured candidate, Curzon, was in the Lords, and that the other major rivals were still allied with Lloyd George.) Mrs Thatcher was helped both by back-bench rejection of Heath ('the peasants' revolt') and the refusal of other contenders to stand against him. She was very much an intruder and forced her way into the leadership.

Mrs Thatcher proved difficult for senior colleagues and opposition leaders to fathom because she was so outspoken in her beliefs. Her ringing denunciations of high public expenditure and high taxation and her defence of the fundamental importance of freedom of choice were deeply felt and frequently stated. She dismissed what she regarded as woolly talk about consensus politics. She was not interested in a compromise with left-wing socialists. Her remarks in a television interview about immigration in 1977 (claiming that people felt 'swamped'), hostility towards trade unions, and support for home ownership or about holding referendums over strikes (again in a television interview), grammar schools, and capital punishment evoked much sympathy among the grass roots. Previous Conservative leaders had lacked such a populist touch. Yet, for this very reason, a number of observers regarded her as abrasive and lacking in compassion.

Politics in Opposition[23]

Despite the ample evidence that Mrs Thatcher was a radical Conservative the image of radicalism was not borne out by her early actions as a party leader. Her Shadow Cabinet continued to be dominated by former colleagues of Mr Heath. Although Robert Carr and Peter Walker were not included, her new appointments did not mark a sharp change in direction. A group that contained such political heavyweights as Mr Whitelaw, Mr Prior, Lord Carrington, Sir Ian Gilmour, and, for a time, Mr Maudling could counter any swing to the right in the Shadow Cabinet. The only obvious supporters of her views were Sir Keith Joseph, John Biffen, and Airey Neave. The official party strategy document, *The Right Approach*, was moderate in tone and content and even earned the approval of Mr Heath at the 1976 Party Conference. In Parliament the front bench abstained on the Labour government's pay limit of six pounds in 1977 and did not oppose the government's rescue of British Leyland and Chrysler; nor did it formally campaign against the closed shop. Mrs Thatcher's caution and her concern for party management were evident in her appointments and in the fine print of Conservative policy statements.

The period of Conservative opposition saw a series of debates, sometimes explicit, sometimes coded, about the nature of true conservatism and what the electorate really wanted. On a number of issues, such as immigration, capital punishment, and trade union power the views of most voters were well to the right of the leadership of both parties. Of course, trade unions had been unpopular for many years and resentment against high taxation was nothing new, any more than the desire for tougher lines on immigration and law and order. But some survey research suggested that the Conservatives would do better to campaign on such populist-authoritarian issues rather than on traditional economic issues.

The emergence of two tendencies in the party became more apparent after 1974. So-called 'drys' favoured monetarist economic policies, cutting public expenditure and taxes and relying on free-market principles. The ideas of Friedman and the rebirth of the quantity theory of money (Chapter 3) were important in providing this wing of the party with a strategy for combating inflation without incomes policy. They also attacked the rise in public spending as a cause of inflation, high taxes, and poor economic performance. Monetarism, in the hands of its Conservative advocates, became part of a general attack on the post-war social democratic Keynesian settlement. So-called 'wets', often associated with Mr Heath, favoured reflating the economy to reduce unemployment, allowing the government to play a more positive role in social reform, and seeking a greater degree of agreement with the trade unions. Of course, a number of Conservatives were not easily located within either camp.

The debate could easily be portrayed as one between left and right-wing versions of Conservatism, or between tough and tender-minded attitudes. The tough-minded became more vocal in opposition. They were opposed to what they regarded as the permissive and egalitarian character of Conservatism of the previous decade and a half. They also believed that a fundamental change was necessary and possible, given political will. We saw in Chapter 4 how Sir Keith Joseph emerged as the chief spokesman of the tough-minded. He drew a distinction between the common ground, the area on which there was general agreement between the parties and which most voters accepted, and the middle ground, a point midway between the policies of the two main parties. He complained that as Labour gave a turn of the ratchet to socialism each time it was in office and this was then accepted by the next Conservative government, so the new position was enshrined as a 'consensus'. But as each successive Labour government moved to the left, enacting collectivist and egalitarian policies, and Conservatives moved to the centre, so the middle ground increasingly moved to the left. This line of analysis certainly impressed Mrs Thatcher and those who looked for a more radical conservatism.

Other self-styled middle of the road voices, predominantly in the upper reaches of the party, doubted the electoral wisdom and political acceptability of these views. They believed that the electorate had become used to high levels of public spending and high rates of tax to provide welfare services and full employment. They also believed that most uncommitted voters were indifferent to ideological appeals and that the constraints on government limited the possibilities for a sharp change in direction. Some came to regard constitutional reform as a way of protecting Conservative values or hindering the damage which a future left-wing Labour government might do. Lord Hailsham, for example, argued eloquently for constitutional reform as a means of limiting the elective dictatorship of Parliament and the power of government.[24] A number of other senior figures favoured proportional representation, in spite of Mrs Thatcher's strong hostility, as a means of heading off a left-wing Labour government. In March 1978 more than a third of Conservative MPs voted for proportional representation for a Welsh Assembly. A party report was issued in September 1978, suggesting the use of referendums in some cases. Others supported a stronger second chamber. One senior 'wet' Shadow Minister commented, 'I think it is more important to keep the country together, to conciliate and have continuity of policies rather than to have "Big C Conservatism".'[25]

Mrs Thatcher, however, showed little interest in institutional reform. Her argument against what she called the 'me tooers' was that if government was overbearing and too socialist then a clear Conservative majority pledged to policies of self-restraint in spending and taxation was the best way to prevent and reverse the ratchet of socialism. The experience of the Lib–Lab pact of 1977 only confirmed her own fears that, with proportional representation, a small Liberal party might gain great blackmail or veto power, and be more disposed to support a Labour than a Conservative government.

The acceptability of the Conservative policies was considerably eased by factors outside the party's control. There was some reaction, not least among Labour ministers, to the steep increases in taxation and public spending between 1974 and 1976. Concern mounted over inflation, which exceeded 25 per cent on an annual basis in June 1975, and over the growth of public spending. In 1975 it was significant that Mr Healey no longer budgeted for full employment, a major breach with the post-war consensus. Another influence was the sudden collapse of sterling in 1976—at one point it fell as low as £1 being worth $1.52—and then the IMF rescue package. From late 1976 a set of new policy instruments was imposed, including tighter controls on public spending. Mr Callaghan reacted to the concern over educational standards by launching a 'debate' on the issue. In voicing concern about the family and mooting the feasibility of appointing a minister for the family he was trying to steal some Conservative clothes.

In three other areas—devolution, incomes policy, and industrial relations—events helped the Conservative party. The party was divided over the devolution proposals for Scotland and Wales, just as Labour was. A majority in the party was opposed to the government's proposals but when the referendums failed to carry the proposals by a sufficient majority in March 1979, devolution died as an issue. The Conservatives were also in a difficult position over the government's incomes policy, which was widely credited with reducing the rate of inflation during 1977 and 1978. But during the Winter of Discontent the government's 5 per cent target for increase in incomes had clearly collapsed. A final problem for the Conservatives had been how to handle the trade unions. Labour's claims that it alone could work with the unions and that a Conservative government would produce confrontation were shown to be hollow. Amid the unruly picketing and well-publicized suffering of the public during January and February of 1979, many union members were contemptuous of agreements and spurned the instructions of their leaders and the exhortations of Labour ministers. Mrs Thatcher was able to capture the political headlines in January with her proposals for trade union reform. These covered suggestions for no-strike agreements in essential industries and taxation of supplementary benefits for strikers.

The Conservative manifesto of 1979 listed its proposals under five main headings. These were:

(1) The control of inflation and trade unions' power.
(2) The restoration of incentives.
(3) Upholding Parliament and the rule of law.
(4) Supporting family life by a more efficient provision of welfare services.
(5) Strengthening defence.

The main proposals for controlling inflation included the strict control of the money supply and a reduction of both the government borrowing requirement and of the state's share of national income. Cuts in public spending were promised in almost every area. The Conservative government would also try to sell their shares in the recently nationalized aerospace, ship-building, and National Freight operations. On trade unions the party proposed three changes 'which must be made at once'. First, it would amend the law to limit the right of secondary picketing. Second, it would provide compensation for workers who had lost their jobs as a result of closed shop and also allow a right of appeal to the courts against exclusion or expulsion from any union. Finally, public money would be provided to finance postal votes for union elections and pre-strike ballots and unions would be required to contribute to the support of strikers. On taxation, the party promised to cut both the top rate of income tax to the European average of 60 per cent and the bottom rate of the tax in order to take the low paid out of the tax net. There would, in addition, be a switch from direct to indirect taxation.

The 1979 Election

The conduct of the general election of 1979 has been explored at length elsewhere.[26] The Conservatives won a decisive victory against a Labour government which never recovered fully from the damage to its authority and popularity inflicted by the winter's industrial disruption. The Conservatives had a majority of seventy seats over Labour and one of forty-three seats over all the other parties combined. The result raised considerable interest, not only because of the new Prime Minister's sex and her personal style, but also her declared determination to reverse so many post-war policies.

The outcome of the 1979 general election was clearly a repudiation of Labour, another stage in the long-term decline of the party (see Chapter 6). But few commentators at the time saw the election as a turning-point. Given Labour's impressive electoral recovery during 1978 it was possible to argue that 1979 had seen a short-term swing towards the Conservative party, largely as a reaction against the Winter of Discontent. Labour was still marginally ahead of the Conservative party when voters were asked to identify with a political party, and the Conservatives were not decisively preferred as the best party to handle issues of prices, unemployment, and trade unions. They were, however, clearly preferred by vote switchers on issues like taxation and law and order. There was also a big swing to the Conservatives among trade union voters and among the skilled working class, who were particularly attracted by the proposals to reduce taxation. Crewe notes that between 1964 and 1979 public opinion shifted to the ideological right and there was a growing correspondence between the views of Conservative supporters and official party policy.[27] The opposite was the case for the Labour party.

Each general election is surrounded by its own myths about the causes of the outcome and about its implications. In particular, there are many myths about election mandates, as victors speedily claim that the voters, having considered alternative party programmes, have now endorsed the winning party's policies. This is a gross simplification of the many reasons which lead people to vote in a particular way. Yet both Mrs Thatcher and Mr Callaghan during the election agreed that the Conservatives wished to depart from many post-war policies.

The Conservatives had won the election, in spite of being led by a right-of-centre leader. Few commentators or colleagues thought that she would prove to be as radical in government as she promised. Expressing the view that the area of choice was limited, Peregrine Worsthorne commented in the *Sunday Telegraph* on 29 April 1979: 'Whatever happens in the election is not going to make much difference. There will be neither revolution nor counter-revolution . . .', and any change will be measured 'in inches not miles'. Sceptics could also note how preceding administrations had been overcome by obstacles and economic failures. But they might also have noted how Macmillan and Heath had managed as Prime Minister to get the party to accept some remarkable leader-initiated switches of policy.

Notes

1. See A. Gamble, *The Conservative Nation* (London, Routledge and Kegan Paul, 1974); Z. Layton-Henry, *Conservative Party Politics* (London, Macmillan, 1980); R. Behrens, *The Conservative Party from Heath to Thatcher* (London, Saxon House, 1980); J. Ramsden, *The Age of Balfour and Baldwin 1902–1940* (London, Longman, 1978); I. Gilmour, *Inside Right* (London, Hutchinson, 1977).
2. But see I. Macleod's review of R. Churchill, *The Fight for the Tory Leadership* (London, Heinemann, 1964), in *The Spectator*, 17 January 1964.
3. E. Nordlinger, *The Working Class Conservatives* (London, MacGibbon & Kee, 1967).
4. (London, Penguin, 1947).
5. E. Nordlinger, *The Working Class Conservatives*, and R. McKenzie and A. Silver, *Angels in Marble* (London, Heinemann, 1968).
6. W. Guttsman, *The British Political Élite* (London, MacGibbon & Kee, 1963); F. Parkin, 'Working Class Conservatism', *British*

Journal of Sociology (1967). R. Miliband, *Capitalist Democracy in Britain* (Oxford University Press, 1983).

7. W. Greenleaf, *The British Political Tradition*, vol. ii (London, Methuen, 1983).
8. P. Norton and A. Aughey, *Conservatives and Conservatism* (London, Temple Smith, 1981).
9. *Modern British Politics* (London, Faber, 1965).
10. H. Macmillan, *Tides of Fortune* (London, Macmillan, 1969), pp. 303–4.
11. *Modern British Politics*, p. 308.
12. S. Finer, H. Berrington, and D. Bartholomew, *Backbench Opinion in the House of Commons* (London, Pergamon, 1962).
13. *The Conservative Party From Heath to Thatcher*, p. 7.
14. R. Butler, *The Art of the Possible* (London, Hamish Hamilton, 1971), pp. 133–4.
15. Private information.
16. *Competition and the Corporate Society* (London, Methuen, 1972), p. 229.
17. *At the end of the Day 1961–63* (London, Macmillan, 1973), p. 37.
18. Cited in A. Gamble, *The Conservative Nation*, p. 108.
19. *Class Conflict and the Industrial Relations Crisis* (London, Heinemann, 1977), p. 67.
20. For unsympathetic accounts of the record see J. Bruce-Gardyne, *Whatever Happened to the Quiet Revolution?* (London, Charles Knight, 1974) and M. Holmes, *Political Pressure and Economic Policy: British Government 1970–4* (London, Butterworth, 1982).
21. See J. Bruce-Gardyne, *Whatever Happened to the Quiet Revolution?*
22. A. King, 'Margaret Thatcher: The Style of a Prime Minister', in A. King (ed.), *The British Prime Minister* (London, Macmillan, 1985), p. 97.
23. This section draws on D. Butler and D. Kavanagh, *The British General Election of 1979* (London, Macmillan, 1980), chap. 4.
24. *The Dilemma of Democracy* (London, Collins, 1978).
25. Interview.
26. See D. Butler and D. Kavanagh, *The British General Election of 1979* (London, Macmillan, 1984) and H. Penniman (ed.), *Britain at the Polls 1979* (London, AEI, 1981).
27. I. Crewe, 'Why the Conservatives Won', in H. Penniman (ed.), *Britain at the Polls 1979*.

8

THE CONSERVATIVE RECORD 1979–1989

THAT governments do make a difference is an important presumption of free competitive elections. Citizens choose between different teams of politicians and whichever one wins enacts the programme on which it campaigned. In fact only about thirty or so states in the world today regularly hold competitive elections and of these only a small number provide an outcome in which one party forms the government.[1] Many changes of government in Britain do not usher in a new set of policies in most departments. Over the long term continuity is more apparent than discontinuity. Historians might say that in the twentieth century in Britain only the reforming administrations of 1906 and 1945 provided a major change in direction to be followed by administrations which consolidated the changes. In the future historians are likely to regard the post-1979 Conservative governments as also having produced a change in direction, one which reversed the direction of previous post-war administrations.

The Impact of Parties

Political scientists have for some time tried to grapple with the question: what difference, if any, does party control of the government make to political outputs? Does a government of the left, for example, produce policy outcomes which favour the working class and organized labour, and shift power from the private to the state sector? Do administrations of the right pursue policies which have the opposite effects? A difficulty with conducting this type of research is controlling for outside factors, for example, international recession, or war, or a disastrous crop failure. And how does one deal with counterfactuals, that is, what might have happened in the absence of the control of government by the particular party? Might another party have followed similar policies? The burden of many studies, subject to such qualifications, is that on balance political factors like party or ideology or electoral competitiveness do not matter much, compared to a country's level of economic development.[2]

In Britain there has developed a learned debate about the impact of party on government. One school, led by S. E. Finer, claims that the parties do make a difference and that, because parties have become more partisan, this is regrettable.[3] These critics point to the discontinuity which followed changes of government in the 1970s in such areas as industrial policy, incomes policy, secondary education, industrial relations and law. They also claim that these discontinuities were aggravated by the more frequent turnover of party control in the government in the 1970s and the influence exercised in both parties by its more partisan groups. Britain, it is suggested, suffers from an *adversary form of politics*, in which a party formulates in opposition—largely for ideological and electorally opportunist reasons—the policies which are then carried into government. A new government's efforts to undo much of its predecessor's work is damaging and disruptive. This style, critics allege, is accentuated by Britain's two-party system and unfair (because disproportional) first past the post electoral system, which produces virtually full power for the government and virtual impotence for the opposition.

A different view has been advanced by Richard Rose.[4] Rose tried to test if there were big shifts in the economy, as posited by the adversary model, or continuities, regardless of changes in party control of government. Evidence of continuity is what Rose calls, unhelpfully perhaps, *a consensus model*. Between 1957 and 1979 he found that on such economic inputs as minimum lending rate, public-sector borrowing requirement, and public spending as a proportion of GDP and on economic outcomes such as inflation, unemployment, and growth, the record showed long-term secular trends, regardless of changes of government.[5] Something stronger than party was clearly at work.

There are many reasons for parties failing to produce intended changes in outcomes. One may point to the ever-present pressures of interest groups and lobbies, international forces beyond the control of a national government, the advice of permanent civil servants inherited commitments in policy and public expenditure, and the flow of so much routine in policy. These are not necessarily forces for continuity but they may act as barriers to government attempts to impose a new line of policy. One might also point to the fear among many politicians of electoral retribution if radical measures turn out to be unpopular. On the other hand critics of the consensus within the Labour and the Conservative parties are more likely to point to the failures of political will, courage, persistence, and policy preparation by their respective leaderships as explanations for the convergence.

Changes in government have led to discontinuities in policy, but trends towards convergence have also occurred, despite the parties in office attempting to carry out different policies. Between 1945 and 1975, for example, there was a change in government policy on prices and incomes every thirteen months on average and between 1958 and 1974 a change in corporation tax every two years on average. In 1970 the new Conservative government reversed its predecessor's policies on the reorganization of secondary education, aid to industry, prices and incomes controls, and pensions. In 1974 Labour introduced new policies for each of these, as well as for the EEC, housing finance, and trade union law. Yet by the end of its term of office each government had moderated its original policies. By 1973 the Conservatives had moved to a statutory prices and incomes policy and massive state intervention in industry; the reorganization of secondary schooling along comprehensive lines proceeded, albeit at a slower pace than under Labour, and the trade union legislation under the Industrial Relations Act was effectively non-operational. By 1978 Britain, under Labour, was still in the EEC, the government had an incomes policy of sorts and had curbed the growth of public expenditure, and private education remained. What is important to note, therefore, is that governments often abruptly or gradually change their policies, generally around the mid-term of Parliament, and that continuities occur in spite of the parties' starting out with different policies. This convergence may have less to do with ideology than with political 'learning' and having to cope with unanticipated events. The adversarial critique may explain the early stages of a government's record but Rose's 'moving consensus' is more relevant to an understanding of the long-term trends.

The Conservative Record

Conservative ministers and many commentators claim that there has been something of a revolution in British politics in the 1980s. Supporters and many critics agree that the old policy agenda has been turned upside down. This chapter considers the record of governments since 1979 in four important areas; welfare, privatization, the economy, and industrial relations. Although there have been more comprehensive examinations of the record of the governments, the above four policy areas are of interest because the Conservative government pledged itself to change the direction of policy in them.[6] They do represent, however, only a part of the record of the government and in some other areas, notably education and central-local government relations, there have been significant changes.

Welfare

The thinking of Conservative critics of the welfare state has been coloured by three broad influences. One is that the cradle to grave provision of welfare, implicit in the Beveridge proposals, has proved to be too expensive and that the demand for welfare has grown faster than has the national income to pay for adequate comprehensive services and benefits. The choice, according to such critics, has been either to increase taxes to cover the public spending for the services—which was contrary to the government's tax-cutting policies—or to reduce the scale of state provision of welfare. The critics argued that some cut-back or restraint in spending or provision was necessary, both to fund tax cuts and allow for the concentration of resources on the most needy. They were *selectivists* rather than *universalists* in their approach to welfare spending.

A second influence has been the judgement that too dominant a role for the state weakens the values of individual self-reliance, family and community solidarity, and private charity. People should be encouraged to make provision for themselves and their families, and state support should provide a safety net for the very poor, without stifling private initiative and self-help. Some Conservatives also suspected that the welfare ethic—by removing the risks and disciplines of the market, providing a cushion for failures, and undermining incentives for the ambitious—weakened enterprise. In 1981 the then Head of Mrs Thatcher's Policy Unit, Ferdinand Mount (author of the book, *The Subversive Family*), set up a Family Policy Group, a committee of Cabinet ministers, with a remit to suggest ways for strengthening family life and promoting a sense of individual responsibility. Among proposals which ministers made were more incentives for mothers to remain at home, education vouchers, curbs on the power of the professions, and greater encouragement for wealth creators.

But offsetting the above views was an awareness among party leaders that many welfare programmes, particularly old age pensions and the National Health Service, are very popular with voters. In 1979 Mrs Thatcher made election pledges to increase old age pensions in line with inflation and to 'protect' the National Health Service, although she made no commitments about other areas. Subsequently she and ministers boasted about their increased spending and claimed that the NHS is 'safe with us'.

TABLE 8.1 *Percentage Change in Programme Expenditure between 1978–1979 and 1988–1989 (in real terms)*

base year 1987–8 (£ million)	% change from 1978–9 to 1988–9
Defence	+18
Foreign Office	− 5
EEC	−40
Agriculture	+13
Trade and Industry	−67
Employment	+73
Transport	−15
DOE/Housing	−74
DOE/others	−11
Home Office	+66
Education	+ 9
Arts	+29
DHSS/Health	+34
DHSS/Soc. Security	+33

Source: Public Expenditure White Paper, Jan. 1989.

For much of the 1980s there was also a growing concern among ministers about the mounting financial costs of the system. Social security spending expanded greatly after 1979 (see Table 8.1) partly because of the unexpected and sharp rise in unemployment and partly because of the increase in people of pensionable age. By 1988–9 the total spending on social security was £47.6 billion or 31 per cent of total public spending, up from 25 per cent in 1978–9. Nearly half (43 per cent) of this total was spent on retirement pensions. The other main categories of social security spending were family allowances (12.5 per cent), supplementary benefits (17.5 per cent), and national insurance spending on those suffering unemployment or industrial injury.

For a government intent on curbing the growth of public spending, social security presented a large target. The Thatcher government, like Mr Heath's in 1970, promised to review existing policies to get greater value for money, eliminate fraud, and reduce waste. The earnings-related supplements to benefits for the unemployed, the sick, and widows, and maternity benefit and injury benefit were abolished. In 1980 the government raised various benefits, including those for the sick and the unemployed, by 5 per cent less than the current rate of inflation, though the abatement was restored in 1983 when the benefits were made subject to tax. In 1981 it cut the link between earnings and unemployment benefits and uprated benefits in line with price rather than wage increases (which have been higher for much of the decade).

Treasury concern about the financial consequences of the long-term impact of the steady rise in welfare expenditure and a slow rate of economic growth remained. The famous 1982 CPRS paper examined the options with a low future rate of economic growth and made radical proposals for cutting the state provision of welfare. The contents of the paper apparently horrified some ministers and, under some pressure, Mrs Thatcher withdrew it from the Cabinet agenda. But not everybody accepted its startling assumptions. For ministers to talk in gloomy terms about the future pressures on public spending risked calling into question the effectiveness of their economic strategy. There was other evidence that, with a moderate (2 to 3 per cent) rate of economic growth, the demographic trends would not impose pressures on the public purse until 2020 and a crisis would occur only if the economy failed to grow by 2 per cent or more.[7]

A major worry to Mrs Thatcher and the Treasury was the future of the State Earnings Related Pensions (SERPS) which had been established by the Labour government in 1975. This had become operational in 1978, although full earnings-related pensions were not due to come into effect until 1998. At the time of its introduction it was hailed as a measure which placed pensions above party politics and was accepted by the Conservative spokesmen of the day. By 1985 approximately half of the work-force was in the state scheme and half in private schemes. The actual pension that was to be provided, about a quarter of average earnings, was rather modest in comparison to the pensions provided in most other Western European states.

In 1984 the Social Services Secretary, Norman Fowler, appointed four review teams covering pensions, housing benefit, supplementary benefits, and child benefits. The results were presented in June 1985 in a Green Paper, The Reform of Social Security (Cmnd. 9517). The Green Paper's major suggestion for saving money in the short term was to cut housing benefit, so reducing expenditure by some £500 million annually. Its most controversial suggestion to cut the financial burden on the state was to phase out SERPS over the following three years. The dependency ratio, that is the number of people of working age per pensioner, would improve from 2.76 in 1981 to 2.99 in 2000. But the ratio would then fall sharply in the second and third decades of the twenty-first century. Mrs Thatcher's own fears were revealed in an interview with the *New York Times* in late January 1984 when she claimed that the long-term social security spending commitments were a financial 'time-bomb' for Britain. By the second and third decades of the twenty-first century the pension commitments would probably involve a major burden upon a smaller number of taxpayers, who might not be prepared to carry it. It was also well known that Mrs Thatcher preferred to have people buying their own private pension scheme; here was an area in which the state might be rolled back.

In Parliament, opposition spokesmen condemned the

proposals as an attack on the welfare state and a break with the consensus. That was expected but much of the reaction from employers, welfare groups, and the pension industry as well as Tory back-benchers was also hostile. Revised proposals were presented in a White Paper in December 1985. These were enacted in 1986 although most did not come into operation until April 1988. The most significant change from the Green Paper was that SERPS would not be abolished but scaled down; the proportion of earnings on which the pension would be based was reduced from 25 to 20 per cent. Although most of the proposed changes in means-tested forms of support were confirmed the review would now realize only marginal financial savings. The government had retreated: radical Thatcherism backed off from fundamental reform of the system.

The appointment of John Moore in 1987 as Secretary of State at the DHSS was seen by some as a signal that the welfare programme would receive a dose of Thatcherism. Moore spoke about his determination to end 'dependency', and drew on the work of the American Charles Murray, who was critical of many US welfare programmes. Yet examination of the aggregate public spending figures on social security hardly justify charges of 'an attack on the welfare state'. On the other hand, uprating many benefits in line with prices rather than earnings means that recipients have not shared as fully as others in the higher living standards achieved in the 1980s. The government's strategy increasingly appears to be one not of reducing expenditure but of containing the rise in costs, targeting benefits on the most needy, and encouraging a redistribution of the burden to the private sector. Child benefit has not been uprated annually since 1987. From the academic year 1990–1 students will no longer be eligible for Housing Benefit, Supplementary Benefit, and Income Support. In 1988 the government established a scheme of Income Support and a Social Fund to replace Supplementary Benefits. Whereas the Supplementary Benefits Scheme had given grants to people to cover essential items, such as cooking utensils, the Fund replaced this with loans. There have also been symbolic measures (such as restoring pay beds in NHS facilities), minor cuts (such as withdrawal of supplementary benefits in vacations for students), some marginal privatization (from 1986 employers were made responsible for paying sickness benefit for the first 28 weeks of an employee's absence from work), tougher criteria (people have to be 'actually seeking work' to be entitled to receive unemployment benefit and since 1988 housing benefit is paid to fewer households), and an increase in the numbers relying on private medical insurance (from 4 to 8 per cent since 1979). But overwhelmingly the state is still the main provider of welfare: Britain has a welfare state.

The story of marginal change is not dissimilar in the health service. For all the talk of 'cuts', state spending on health has increased by nearly a third in real terms since 1979; and as a share of total public spending it has moved from 14 per cent in 1978–9 to 16 per cent in 1988–9. However an increase is needed to take account of the growing number of elderly (heavy users of the NHS) and developments in health care technology, and increased spending has been more modest in some parts of the service, notably hospitals. British health spending of 6 per cent of GDP is a smaller proportion than most other western states. The government claims that total spending is larger in other western states because many have a larger private sector; in the United Kingdom it is just 0.8 per cent of total spending. It wishes to promote more private health schemes although it has made clear that the NHS will remain a largely state-financed service. The 1989 review of the NHS allowed tax relief for private health insurance for the over-60s but the rejection of other means of finance rules out any fundamental change in the near future. The review abolished local authority representation in health service bodies, gave a greater role to the Audit Commission to ensure value for money, gave doctors budgets and linked a greater share of their salaries to the number of patients treated, and held out the possibility of hospitals 'opting out' and becoming self-governing. But these largely managerial changes hardly amounted to a fundamental review; rather they were concerned to promote competition and value for money.

Over time the rhetoric of government spokesmen has changed to boasting of how much more manpower (doctors and nurses) and money is devoted to the NHS, compared to 1979. The government's early arguments that 'we can't afford it' are less persuasive with the budget surplus accumulated in 1987 and 1988 and claims that tax cuts are better than greater state spending on social and health services have found fewer takers among the public (see below p. 297).

Privatization

The impact of the Thatcher government's new thinking has most obviously been felt in the privatization programme for the state-owned and state-provided services. The programme covers both denationalization (or the sale of assets and shares owned by the state) and liberalization (or the relaxation or abolition of a service's statutory monopoly). Perhaps in no other area of policy has the discontinuity between the present and previous governments been so marked. Before 1979 the Conservative party had effectively acquiesced in most of the public ownership measures of earlier Labour governments. Previous Conservative acts of denationalization, such as iron and steel after 1951 (later renationalized by Labour in 1967) and the sale of the Thomas Cook travel agency and Carlisle public houses under Mr Heath hardly refuted the broad continuity of post-war policy or Sir Keith Joseph's complaints about the ratchet effect in this area. The privatization policy was consistent with the Prime Minister's beliefs that there should be a major diminution in state intervention in the economy and greater reliance on the free market. The 1979 Conservative election manifesto claimed: 'The balance of our society has been increasingly tilted in favour of the state at the expense of individual freedom . . . this election may be the last chance we have to reverse that process.' Mrs Thatcher also believed that the publicly owned industries were inefficient and an obstacle to the creation of a more dynamic and adaptive economy.

Apart from the Labour party and public-sector trade unions, the nationalized industries had few friends by 1979. Opinion surveys showed that public ownership won few votes for Labour and that even many loyal supporters disapproved of the party's commitment to further nationalization. Yet a large public sector appeared to be an almost inevitable part of the modern economy. Even when industries were not state owned there were various other instruments by which government actually exercised a large measure of control over the private sector. Government was a major actor through its policies in the public sector, periodic controls on prices and incomes, legislation on industrial relations, policies for regional aid, and the location of firms and trade. The Heath government had taken statutory controls over prices, incomes, and dividends and its 1972 Industry Act permitted large-scale state intervention in industry. By 1979, after another spell of Labour government the nationalized industries accounted for 10 per cent of GDP, 15 per cent of national investment, and 8 per cent of employment. The public sector itself accounted for some 30 per cent of employment. Attempts to roll back the state in industry (as well as welfare) seemed a hopeless task.

Within the Conservative party, particularly among local activists and MPs, there had long been numerous critics of the nationalized industries and advocates of a more market-orientated economy. A more political concern was the legacy of the 1970–4 government's experience with incomes policies and public-sector wages. Both free collective bargaining and incomes policies were difficult to operate in a situation where there were powerful public-sector trade unions, conducting centralized wage bargaining and virtually monopolizing key services. In industries like coal, railways, and power the unions were in a very strong bargaining position. In the mid-1960s wage bargaining in the public sector became more politicized and produced frequent confrontations with the central government, regardless of whether it had an incomes policy or not. The leaked findings during 1978 of an internal report for the Conservative party noted that in the case of strikes in key services and industries the government would simply have to pay up. The party did not want a repeat of the disastrous 'Who Governs?' election of 1974.

Many Conservative spokesmen paint a view of two sectors. On the one hand, a private sector which allowed free entry of producers, was subject to competition, sought profits, and had to respond to consumers or else go bankrupt. On the other side was the public sector, enjoying a monopolistic and low-risk position, responding to political rather than consumer preferences, and pre-empting investment funds from more productive areas. Since 1979 the programme of privatization has been defended with a mixture of ideological and pragmatic arguments.[8] One is that such policies produce an extension of economic freedom and, thereby, political freedom. The nationalized industries, it is claimed, deprive the consumer of choice and limit the taxpayer's freedom to spend his money as he chooses because of the industry's monopoly position and reliance on state subsidies. Critics also pointed to the lack of competition, absence of commercial disciplines (although ministers could set external finance limits and financial 'targets' for return on capital), and threat of bankruptcy. Another criticism was of the poor performance of the nationalized industries, particularly the low pre-tax real rates of return on the capital employed.[9]

In defence of the industries, however, one has to make an allowance for the social costs of policies (such as maintaining an uneconomic rail service for a community or loss-making plants in areas of high unemployment) and for decisions on pricing, investment, and employment which are shaped in part by the political calculations of the government of the day. At times (for example, under Mr Heath) governments have abandoned economic pricing as part of an anti-inflation programme; at other times, notably over the price increases in gas between 1980 and 1982 when the Thatcher government each year raised prices 30 per cent above the current rate of inflation, above 'economic' pricing levels and what the board wished. The vagueness of guidelines and frequent and inconsistent ministerial interventions were frustrating for management of nationalized boards.

The government therefore intially turned to privatization in large part as an answer to the 'problem' of the nationalized industries. One minister commented in March 1982: 'Look, we're bloody fed up with them. They make huge losses, they have bolshie unions, they are feather-bedded. It seems almost impossible to do anything with them; therefore, the view has grown, get rid of them.'[10] 'Getting rid' meant selling to the private sector. Earlier plans for tighter financial control and greater independence for management had not been successful. There did not seem to be a half-way house between public and private ownership. Some industries acquired chairmen with a reputation for cost-cutting and a tough stand on labour relations. Ian MacGregor was appointed chairman of British Steel in 1980 and of British Coal in 1983, John King of British Airways, and Michael Edwardes of British Leyland. The remit of the last two included plans to prepare BA and BL for a return to the private sector.

The privatization programme has been recognized as a major break with the mixed economy consensus. The verdict of one admiring authority on the government's performance is that, 'In doing so, it has achieved the largest transfer of property since the dissolution of the monasteries under Henry VIII, a transfer from the state to its citizens.'[11] What is remarkable is how little political opposition there has been to the programme to date. In spite of complaints by leaders of trade unions many of their members bought up the shares. There were few strikes and in Parliament Labour's opposition was usually focused more on the 'giveaway' price at which shares were sold rather than on the principle of privatization. Mr Kinnock has suggested that renationalization would not be a high priority for a future Labour government.

The government probably went further down this road than it had originally envisaged. The 1979 manifesto had promised to restore to the private sector the recently nationalized aerospace and shipbuilding industries, sell off the shares in the National Enterprise Board and National Freight Corporation, and review the role of BNOC. But by 1983 sales had been arranged for part of Britoil, Associated British Ports, Amersham International, shares in BP, and the NEB's shareholdings (Table 8.2). These sales realized nearly £1.8 billion. The programme then developed a momentum of its own. In the 1983 Parliament, Sealink, Jaguar, British Telecom, British Gas, and British Airways were all privatized.

In 1985–6 alone, the sale of the rest of Britoil, receipt of the second British Telecom payment, and the sale of British Airways raised over £2.5 billion. By 1988 the programme had shifted 600,000 jobs from the public to the private sector and reduced the nationalized industries' share of GDP to 6.5 per cent. The latter will have lost the national telephone system, gas and steel industries, and the country's largest airline. By 1990, it is planned that electricity and water, the core utilities, will also have been sold off. Ministers claim that a fourth-term

TABLE 8.2 *Privatization in Britain 1979–1989*

Company	Business	Date of Sale
British Petroleum	Oil	1979
		1981
		1983
British Aerospace	Aerospace	1981
British Sugar Corporation	Sugar refiner	1981
Cable & Wireless	Telecommunications	1981
		1983
Amersham International	Radio-chemicals	1982
National Freight Co.	Road Haulage	1982
Britoil	Oil	1982
Associated British Ports	Seaports	1983
		1984
International Aeradio	Aviation Communications	1983
British Rail Hotels	Hotels	1983
British Gas onshore Oil Assets (Wytch Farm)	Oil	1984
Enterprise Oil	Oil	1984
Sealink	Harbour and Ferry	1984
Jaguar	Cars	1984
British Telecom	Telecommunications	1984
British Technology Group	miscellaneous	
British Gas	Gas	1986
Vickers Shipbuilding & Engineering Ltd	Engineering	1986
Royal Ordnance Co.	Arms	1987
British Airways	Airline	1987
British Airports Authority	Airports	1987
Rolls Royce	Engineering	1987
British Steel	Steel	1988
Rover Group	Cars	1988

Thatcher government would tackle the final two monuments of post-war nationalization—coal and railways.

Over time the original emphasis of the programme has broadened from increasing liberalization and competition to raising money for current spending.[12] In all, some £24 billion had been raised from sales by the end of 1988. Sales of state assets helped the government to reduce the public sector borrowing requirement. There are, of course, various techniques of privatization. In one case the government simply sold the whole enterprise to private investors such as Amersham International and the Atomic Energy Authority. In another it sold a part of the whole enterprise by converting the public corporation into Companies Act companies and selling half of the shares: it did this, for example, by reducing its stock in BP from 51 to 46 per cent in 1979, then to 39 per cent in 1981, and again to 31.7 per cent in 1984. A third technique was to sell off a proportion of the whole operation, as it did with British Telecom, British Gas, British Aerospace, and Britoil; and British Rail was made to sell off its hotels. A fourth was to sell the shares to the work-force, as happened with the National Freight Corporation. In some key sales, such as British Gas and British Telecom, the government retained a share of the equity to protect the national interest. Yet another method was to liberalize entry or remove restrictions on firms wanting to compete in the provision of a service in the public sector. A number of local authorities, for example, put out refuse collection to private tender.

The privatization programme also covers schemes to promote deregulation and competition in the economy. Within the state sector, the statutory public monopoly of electricity supply and express delivery service has been ended. The 1980 Transport Act, by relaxing controls over licensing and fares, has increased competition on long-distance coach routes. In January 1986 any operator was automatically licensed to run on any route outside London. Between 1981 and 1983 the government created twenty-four so-called enterprise zones in economically depressed areas. This provided a package of financial incentives and exemptions from various laws and regulations to encourage businesses to set up in the zones. In 1984 six freeports were established in areas near airports or the sea at Liverpool, Birmingham, Cardiff, Southampton, Belfast, and Prestwick.

The policy of increasing the role of the market sector has extended to contracting out services in the National Health Service and local government and the removal of the monopoly powers of solicitors (over house sale conveyancing), or barristers appearing in the high court, opticians, and financial services via Big Bang in 1986.

There is no doubt about the political and economic significance of this programme. It has redrawn the boundaries between the public and private sectors in favour of the latter and raised substantial sums for the Treasury. Although rarely stated by ministers as a reason for privatization, the funds could be used to finance tax cuts and/or current spending. In other words, it was a way of easing the dilemma of choosing between tax cuts or more public spending. Some of the claims made by government defenders of the programme are, however, questionable. The sales have done little to increase competition: BT and British Gas retain their monopolistic position, although their pricing policies are subject to review by government appointed regulatory bodies. Dunleavy and Rhodes have noted that privatization in these sections has, ironically, involved an *increase* in regulation or re-regulation.[13] The electricity supply industry is to be broken up and sold. Some of the more spectacular productivity gains (e.g. British Steel) have occurred prior to privatization. Some observers also calculate that, if one takes account of the profits forgone from these enterprises in future years and sets them against the revenues raised now, the government may have sold the assets at too low a price.

Separate from but related to the programme of privatization of state-owned industries and deregulation was the scheme for the sale of local council houses. The programme advances the traditional Tory theme of a 'property-owning democracy' and was started by the Heath government. It was outlined in the Housing Act (1980), which gave council tenants with three years' residence the right to buy their properties at a substantial discount. The discounts were extended by another Housing Act (1984). By December 1988 over a million had purchased their properties.

The Economy

Arresting the cycle of the country's relative economic decline was the self-proclaimed major task of the Thatcher revolution. Conservative leaders claimed that by curbing the growth in public spending they would create the headroom for tax reductions which in turn would provide incentives and liberate the entrepreneurial energies of the British people. By strictly controlling the money supply they would restore sound finance and squeeze inflation out of the system. The government would not intervene in private sector wage bargaining but intended that its own example in the public sector plus strict control of the money supply would encourage responsible wage bargaining. Ministers also hoped that employers would cut out over-manning and root out inefficiency, even if this led to a temporary surge in unemployment. Reducing the size of the public sector and number of regulations that hampered business would create the opportunity for the emergence of a more market-orientated economy. Reforms of the internal practices of trade unions and legal changes in their status would provide opportunities for managers to regain authority and for the emergence of more 'responsible' and less 'political' trade unions. Finally, by reducing the budget deficit and government borrowing they hoped to bring interest rates down.

The ultimate objectives of this strategy were of course no different from those of previous post-war governments. All governments have wanted to strike a balance between the economic aims of low inflation, rapid economic growth, a surplus on the balance of payments, and full employment. But, increasingly in the 1970s, governments were failing to achieve the objectives (see above, Chapter 6). It is important to realize, however, that what some commentators have called the 'new realism' did not start in 1979. Mr Callaghan's famous speech to the 1976 Labour party conference (subsequently cited in many Conservative party publications) admitted that governments could not spend their way into full employment; that way only led to more inflation and eventually more unemployment. A squeeze on public spending, adoption of money supply targets, privatization (sale of BP shares in 1976), and the priority accorded to fighting inflation, even at a time of historically high unemployment, were all in place under Mr Callaghan.

(i) Inflation

Of all the economic variables inflation is the one that the Thatcher government has said it is able to control. According to the Chancellor, Nigel Lawson, in his Mais Lecture in June 1984: 'It is the conquest of inflation, not the pursuit of growth and employment, which is or should be the objective of macro-economic policy.' In many respects the 'New Conservatism' in Britain and the United States was born out of a reaction to the high inflation of the mid-1970s. In the 1950s and 1960s the annual rate of British price rises was modest by comparison with what was to follow in the 1970s. Yet the average rate in Britain (4.1 per cent) between 1961 and 1970 was still higher than that in OECD countries (3.3 per cent). In the years 1971–80 the gap widened to annual averages of 13.7 per cent and 9.0 per cent respectively.

Reducing inflation was, until 1989, the government's greatest achievement on the economic front. The increase in oil prices combined with the 1979 June budget's increase from 8 to 15 per cent in VAT on many goods and high wage rises helped to push inflation up to 21 per cent in May 1980. In the second half of 1982, however, Britain's inflation rate fell to the OECD average of 5 per cent and then steadily declined to 2.5 per cent in July 1986, the lowest figure since 1967. The reduction was helped by the strength of sterling which became increasingly attractive as an oil-backed currency. Although exporters fairly complained that this made their goods less competitive in price, it helped to lower the cost of imports and thus the prices of foreign goods sold in Britain. In 1987 Conservative ministers boasted of their success while still claiming that the conquest of inflation remained the first objective of policy. According to the election manifesto: 'There is no better yardstick of a party's fitness to govern than its attitude to inflation. Nothing is so politically immoral as a party that ignores that yardstick.'

After 1974 a growing number of Conservatives increasingly fastened upon the failure to control money sterling supply, M3 (notes and coins in circulation, with all deposits held by UK residents), as the main cause of inflation. Monetarists claimed that there was a variable time-lag (of between eighteen months and two years) between an increase in money supply and the consequent price inflation and that the huge expansion in money supply (28 per cent in 1972, 27 per cent in 1973) caused the soaring inflation of 1974–5. Politics joined with economics in the decline of incomes policy and the rise of monetarism as the appropriate weapon in the fight against inflation. Monetarists argued that as long as the government refused to increase the money supply, even if unemployment was increasing, there could be no inflation. Controlling and adhering to announced targets for money supply became a test of the government's will. The announcement in the Medium Term Financial

Strategy (MTFS) in the 1980 budget of monetary targets—which would fall year by year (see Table 8.3)—was also regarded as a means of lowering the expectations of wage bargainers. The 'rational expectations' school of monetarists argued that expectations of lower inflation would lead to a slow-down in the rise of wages, costs, and prices.

Yet the government's monetary policy was hardly successful in reducing inflation. It had many difficulties in controlling M3, which proved to be an unreliable barometer. An influential report by Professor Jurg Niehans in February 1981 argued that monetary policy was in fact excessively tight because the high interest rates were attracting foreign money to London and maintaining a high pound.[14] Difficulties in exercising control were also increased by the civil service dispute during 1980 and the increase in bank lending for mortgages. Another measure, of 'narrow' money, was Mo, which covered cash, the banks' till money, and the Bank of England's operational cash, and the Chancellor set a target for this as well. Table 8.3 shows that the government regularly exceeded its monetary targets. The growth of M3 for 1980–1 (19.5 per cent) easily exceeded the planned 7–11 per cent, and between March 1980 and March 1984 M3 grew by 70 per cent, against its target 46 per cent. By 1982 government spokesmen spoke less about money supply and were more concerned about monetary growth and the exchange rate. In 1984 the target was concentrated on Mo, the narrower measure. In the Lord Mayor's speech in London in November 1985 Nigel Lawson formally abandoned M3 and since 1987 gave more emphasis to broad money or M4 (which is M3 plus building society liabilities) though not setting a target range. Monetarism was effectively abandoned. Management of interest rates and the exchange rate were seen as more effective influences. By 1989 monetarism was attracting more support, as Britain had one of the highest inflation figures in the European Community (8 per cent in April 1989). M4 had expanded rapidly between 1986 and 1988.

(ii) Public spending

The government only gradually reduced the growth rate of public spending. Control of spending was crucial to the government's strategy because it wanted to cut governmental borrowing and taxes. Yet in spite of several Cabinet meetings to make reductions, total spending actually increased in real terms and as a proportion of GDP. Ironically, the government was widely criticized for making 'cuts'. Cutting public spending has always been a politically divisive process for a government, as ministers in the major spending departments battle to protect their budgets. In the post-war period the only previous cuts in public spending as a proportion of GDP were achieved by Labour in 1975–6 and 1977–8.

The Conservatives' original goal in opposition was to stabilize the total public spending figure in real terms at its 1977 level and for that total to fall as a share of GDP as the economy grew. The MTFS made a deliberately modest assessment of future rates of economic growth (1 per cent per annum 1980–3). In his 1979 budget Sir Geoffrey Howe claimed that in the past public spending had been based on falsely optimistic expectations about economic resources: 'It is this falsely reassuring belief that somehow the resources will be found to permit an uninterrupted expansion of public expenditure that this government challenges. In planning public expenditure it is better to be prudent and make a deliberately cautious assumption on the growth of future resources.' The MTFS in 1980 planned for a reduction in public spending of 5 per cent over the next four years; there was in fact an 8 per cent growth in real terms. Given the party's pledges on pensions, defence, law and order, and other statutory commitments, particularly on social security, this meant that spending cuts would have to fall in a few areas, notably housing. The increase was in large part caused by the

TABLE 8.3 *The Medium Term Financial Strategy: Projections and Outturns* (Money Supply: sterling M3)

	1979–80	1980–1	1981–2	1982–3	1983–4	1984–5
June 1979	7–11					
March 1980		7–11	6–10	5–9	4–8	
March 1981			6–10	5–9	4–8	
March 1982				8–12	7–11	6–10
March 1983					7–11	6–10
March 1984						6–10
Outturn	11.2	19.4	12.8	11.2	10.1	9.8

rising costs of social security, itself a consequence of the steep rise in unemployment, and the NATO commitment to boost defence spending by an extra 3 per cent per annum in real terms. The failure to reverse the spending growth was a disappointing outcome for the advocates of cuts in the many bruising Cabinet battles over public spending between 1979 and 1982.

There have been marked shifts in spending between programmes since 1979. It is worth noting, however, that the spending priorities have not been radically dissimilar from those planned by the outgoing Labour government, which also planned for cuts in spending on education and housing and an expansion in spending on defence, law and order, and social security. Table 8.1 above shows that between 1978–9 and 1988–9 there was a major and planned increase in real terms in spending on defence and law and order; spending on housing and industry were heavily cut, while on education (helped by a fall in numbers of school children) and on transport it stayed fairly constant.

For all the concern over the level of public spending the British share as a proportion of GDP in 1979 was only at the mid-point of most western states. The Treasury fought and lost many spending battles in the first Thatcher administration; public spending increased from 39.5 per cent as a proportion of GDP, which the government inherited in 1979, to 42.5 per cent in 1984.

The record, until the middle of the decade, disappointed a number of Mrs Thatcher's more ambitious supporters. According to *The Times*'s leader 'What is Thatcherism Now?' on 16 January 1984:

> But none of this is the essence of Mrs Thatcher's problem which is rather that, on the essential questions of economic management, the government gives the impression either of having lost its momentum or of having decided that it must settle for something well short of what Mrs Thatcher seemed to promise in the way of reducing the size of the public sector and also the burden of taxation.

The reduction in unemployment and continued economic growth gradually helped the ratio to fall. But it was only in 1988–9 that the ratio (excluding privatization receipts) fell back to that of ten years earlier.

(iii) Taxes

In line with the government's objective there was a reduction in the top (from 83 to 60 per cent) and standard rates of income tax (from 33 to 30 per cent) in the 1979 budget, and a shift to indirect taxes. In that budget speech Sir Geoffrey Howe stated: 'Our long-term aim should surely be to reduce the basic rate of income tax to no more than 25 per cent.' The same budget increased VAT from 8 to 15 per cent. Later budgets reduced the rates of income tax further; in 1988 the top rate was cut to 40 per cent and the basic rate to 25 per cent. However, claims that the government has reduced the 'tax burden' need to be heavily qualified. The main shift has been from direct to indirect taxes. In 1979 30 per cent of government revenue was raised by income tax; by 1989 the figure was 24 per cent. Less 'visible' increases in VAT and employers' national insurance contributions offset the well-publicized cuts in income tax. Overall total taxation, including corporate taxes, increased as a share of GDP from 33.1 per cent to 37.6 per cent in 1988–9. The rise is largely accounted for by rising incomes and profits which bring more taxpayers and companies into higher tax brackets. In total the tax position is not much different from 1979. Britain's share of taxation as a proportion of GDP has remained in the middle for the range of OECD states.

(iv) Distribution

The final report of the Royal Commission on the Distribution of Income and Wealth drew attention to the stability in the distribution of income in the post-war period.[5] Some redistribution had occurred between 1938 and 1949, but if one ignores the top 1 per cent of income earners, there had been little change up to 1979. Under the Conservative government, as noted, the tax system became less progressive, with the major reductions being made at the top end of the tax range and there were increases in national insurance and the non-progressive indirect taxes. The tax cuts for the top income earners, the increase in share prices over the decade, and the effects of the recession have sharpened economic differentials in the Thatcher period. Although the Commission was dissolved in 1979, the 1985 volume of Inland Revenue Annual Statistics found that between 1979 and 1983 the most wealthy 50 per cent of the adult population increased its share of total marketable wealth from 79–83 per cent to 80–4 per cent. In the period 1979–87 the bottom 10 per cent of earners made net gains of 5 per cent, the top 10 per cent gains of 28 per cent.[6]

Analysis by the Institute for Fiscal Studies shows that richer households have benefited more than poorer ones from the government's tax and benefit changes. Although most have gained, half of the total giveaway in terms of benefits and tax reductions has gone to the richest 10 per cent of households. The single unemployed living on their own are the only group who have lost significantly from the changes—by an average of £2.16 a week.[7]

(v) Unemployment

The lifetime of the Thatcher governments has been dominated by large-scale unemployment. The government inherited a figure of 5.4 per cent (1.2 million) for unemployment which by November 1983 had reached 12.7 per cent (3 million). The rise in unemployment frustrated the government's spending plans, because of the rising costs of social security; it also undermined the credibility of the government's claims that its economic strategy was working and that economic recovery was around the corner. Industrial output fell by over 11 per cent in the course of the 1979–83 Parliament, a performance worse than in other western states which also suffered from the recession. The main damage was done in the first twelve months, when GDP fell by 3 per cent, industrial production by 9 per cent, and unemployment doubled to 2.5 million. The government redefined the categories of unemployed in 1982 and 1983, for example, excluding men over 60, counting only those claiming benefits instead of those registered as unemployed (this change excluded mainly married women seeking work, but not entitled to benefit), and various training and employment schemes for young people kept perhaps another 600,000 off the unemployment list. The figure peaked at 3.2 million, excluding school-leavers in June 1985. Thereafter it steadily declined, month by month and in January 1989 fell below two million (Table 8.4).

TABLE 8.4 *Unemployment: Percentage of Total Labour Force*

Year	UK DE	UK OECD	EEC	USA	Japan
1973	2.1	3.0	2.9	4.8	1.3
1976	4.5	5.6	5.0	7.6	2.0
1977	4.8	6.1	5.4	6.9	2.0
1979	4.3	5.0	5.7	5.8	2.1
1982	9.8	11.3	9.6	9.5	2.4
1986	11.4	11.2	10.9	6.9	2.8
1987	10.2	10.3	10.7	6.1	2.8
1988	8.1	8.3	10.2	5.4	2.5

Notes: OECD figures, partially standardized. EEC figures cover Belgium, France, Germany, Netherlands, Spain, and UK.

Source: S. Brittan, 'The Government's Economic Policy', in D. Kavanagh and A. Seldon (eds.) *The Thatcher Effect* (Oxford, University Press, 1989), p. 23.

(vi) Economic growth

Backed by North Sea oil the pound was an attractive currency on the international money market. In spite of complaints by business about the pound's appreciation (in May 1979 its effective exchange rate was 81.3 per cent and in January 1982 it was 92 per cent of its 1975 rate) the government came to rely on a strong pound as part of its anti-inflation strategy. Contrary to declarations that there was no government policy for sterling and that it would be left to the markets to decide, by late 1981 the government intervened to reverse a fall in the exchange rate, and raised base rate to 16 per cent. It did the same again in January 1985; when the pound fell to $1.12 banks were instructed to raise interest rates by 1.5 per cent. The record high interest rates were attractive to holders of sterling even though they were also damaging to business.

The exchange rate policy was highly damaging to manufacturing. Alt has blamed oil for this appreciation, while other economists have blamed the government's tight money policies and high interest rates.[18] Industrial output fell by 11 per cent between the second quarter of 1979 and the last quarter of 1982, and between 1979 and 1985 industry shed 1.5 million workers. The balance of trade in manufacturing swung from a surplus of £2.75 billion in 1979 to a deficit in 1984 of £3.75 billion and the years 1983 and 1984 were the first time in history that Britain's trade balance was in the red on manufactures. Non-oil imports increased by an average of 4.5 per cent per annum from 1979, while non-oil exports increased by 1 per cent per annum. Only in 1984 did non-oil output recover to its 1979 level. Some decline of the manufacturing trade balance was probably inevitable, because of the advent of North Sea oil at a time of floating exchange rates and the increased appreciation of sterling. But the loss of international competitiveness was much bigger in Britain between 1979 and 1981 than in other western states.

There have been claims of a productivity miracle in British manufacturing and that the 'British disease' has been banished. There are problems in assessing these claims, in part because of disputes over the starting date—1979, when Mrs Thatcher came to office, or 1981, the low point of the recession—and whether comparison is made with the 1970s or 1960s.[19] Over the period 1979–88 manufacturing output per head increased by 4.2 per cent, faster than the 1960s. But for the whole economy the increase is only 1.9 per cent, better than the 1970s but lower than the 1960s. At last British output per head is now up to the European average, largely because the latter has slowed down so much. But Britain still has a huge gap to close on her major competitors.

In defence of their economic record, government spokesmen have claimed that from the second quarter of 1983 the numbers in employment increased and that more jobs were being created in Britain than in the rest of the EEC states taken together. At first they pinned most of the blame for the loss of jobs on the trade unions and the international recession, and then on high wage settlements. Lower wage settlements, claimed the Treasury and the Department of Employment, would help to price people back into jobs. But real wages steadily increased in spite of recession and high unemployment. As the government exceeded its monetary targets and unemployment soared, so critics in 1980 and 1981 waited for a U-turn, or a change in economic strategy. But Mrs Thatcher was adamant about not reflating and this resolution, in spite of the sharply increased unemployment, was important in sustaining the government's credibility. Particularly important in showing that the MTFS was serious was the 1981 budget when the Chancellor increased taxes by £5 billion, or 2 per cent of GNP, to protect his PSBR. This was a major squeeze at a time of recession. Maintaining the policy in face of previously unthinkable levels of unemployment became a test of 'resolute leadership'. In the first four years it nearly halved the PSBR, or borrowing requirement, from 5 per cent of GDP in 1978–9 to 2.7 per cent in 1982–3 and Britain had a markedly tighter fiscal stance than her Western European neighbours. In spite of the recession the PSBR continued to fall, though not as fast as had been forecast.

Industrial Relations

The mood of 'time for a change' was certainly felt in the area of industrial relations, particularly in relations between the government and trade unions. Formal incomes policies and the 'social contract' approach were hardly politically credible after the 1979 winter's disruption, and many features of the trade unions were as unpopular as ever (such as the closed shop, unofficial strikes, close relations with the Labour party, and mass picketing). Under Mrs Thatcher, Conservatives were more outspoken in blaming the unions for low productivity, overmanning, restrictive practices, and strikes, particularly in the public sector where the unions were often able to exploit their monopoly position.

Because the government abandoned any formal incomes policy there was less call for it to maintain close relations with union leaders. The government hoped that strict control of the money supply and the government's stated unwillingness to rescue financially troubled firms would help to inculcate a more responsible outlook among wage bargainers. In the public sector the government would give each department a cash-limited budget, including a 'factor' for pay; ministers claimed that workers, by pushing for 'excessive' wage rises which employers were unable to pass on in price increases, could price themselves out of jobs. For the first two years, however, this strategy lay in ruins because of the consequences of accepting the 'catching-up' comparability awards of the Clegg Commission for public-sector pay.

The Clegg Commission was abolished in 1980 and eventually the government managed to squeeze down pay in the public sector through its use of cash limits. Rising unemployment and fear of job losses weakened the appeal of militants in the unions. The government was the first to do without an incomes policy, and the absence of 'norms' or 'targets' for incomes in nearly twenty years helped to take the issue of wages out of politics. On the grounds of preventing young workers' wages rising to a level which might price them out of work it scrapped the 1946 Fair Wages Resolution, which required compliance with minimum wage standards for firms trying to secure government and other contracts. In 1985 it also reduced the number of Wages Councils affecting the pay of young workers.

The government also took the view that the law had a role to play in redressing the balance in bargaining between employers and unions. In retrospect many Conservatives felt that the ill-fated 1971 Industrial Relations Act had been too ambitious and that the ground had not been adequately prepared for such a sweeping measure. Before the 1979 election the party took advantage of the Winter of Discontent to toughen its manifesto proposals, promising to curtail flying pickets, or picketing away from the pickets' place of work, and provide funds for pre-strike ballots of union members and for the election of union officials. The 'step by step' approach was reflected in the two Employment Acts of 1980 and 1982. These restricted lawful picketing to the pickets' own place of work and removed the unions' legal immunities from civil actions, so making them liable for damages up to a certain limit where they were responsible for unlawful industrial action. The 1980 Act removed the provisions of the Employment Protection Act which affected small businesses, by increasing the qualifying period for complaints of unfair dismissal. The same Act also provided compensation for people unreasonably excluded or expelled from a union in a closed shop, and required that new closed shops in future should be approved by four-fifths of the workforce. These provisions were strengthened in the 1982 Act which provided that no closed shop should be enforceable unless it had been approved by a clear majority of employees voting in a secret ballot. It made trade unions (rather than just individual organizers of strikes) liable for damages arising from unlawful industrial actions. It also gave employers legal remedies against industrial action in which no dispute exists between employers and their own employees or which is not wholly or mainly about employment matters.

The 1984 Trade Union Act provided government funds for the regular election of trade union officials and for elections on whether unions should have a political fund. It also required that industrial action be supported by union members in a pre-strike ballot if the union was to be exempt from civil action for damages arising from the dispute.

A third Employment Act (1988) removed various closed shop immunities, extended the number of union officials to be directly elected, now by postal ballot, and gave legal protection to members who suffered 'unjustified' discipline by a union (e.g. for refusing to support industrial action).

The government successfully pursued a 'hands-off' policy in industrial disputes. In the first term of office, strikes at British Leyland, British Steel, and British Rail, and lengthy disruption by civil servants and health service workers failed to move ministers. Workers at British Leyland and in the coal industry rejected the advice of union leaders to strike over wages or job losses. Perhaps the only two successful threats of action were by the water-workers in 1983 and the NUM over the Coal Board's threat to close uneconomic pits in January 1981. The twelve-month strike by the National Union of Mineworkers against pit closures ended in April 1985 in failure and with the NUM split. The NUM's main trouble with the courts arose over actions brought by its own members on the grounds that the union had broken its own rules (over a pre-strike ballot) and for contempt.

The government managed to make its legislation 'stick', in contrast to Mr Heath's experience, and has refuted the old 'liberal' view that the courts did not have a useful role to play in industrial relations. The idea of unions balloting their members before taking industrial action took hold. But most ballots endorsed proposals for industrial action and all of the union ballots held in 1985 on maintaining a political fund overwhelmingly endorsed the idea. Although it was TUC policy for member unions to refuse government money for ballots, a number of unions defied it. In 1985 the second largest union, the AUEW, was faced with the threat of expulsion from the TUC for breaking the policy. The Electricians Union was prepared to follow the AUEW and, already faced with the breakaway Democratic Miners' Union in Nottingham, the TUC was forced to find a face-saving formula to avert a possible split. The electricians also incurred the displeasure of fellow trade union leaders by making no-strike agreements with employers, arranging private health care facilities for workers, and then secretly negotiating with the Murdoch press for jobs at the new Wapping plant. They were expelled from the TUC in 1988 for the former.

In many respects the Thatcher years have been depressing for the trade unions. The setbacks include mass unemployment, decline of Labour, loss of members, privatization of parts of the public sector, cash limits in much of the public sector, which limited opportunities for bargaining, government initiated incursions into their internal affairs, and minimal access to Whitehall. In July 1987 Mr Lawson unilaterally announced a reduction in the meetings of the tripartite NEDC from ten to four meetings annually and in January 1989 the Secretary of State for Employment announced that the TUC would no longer be the sole body which could nominate trade unionists to the Training Commission and other official bodies. Bargaining rights were weakened. The Pay Research Unit for the civil service was abolished, as was pay bargaining for schoolteachers, and the role of Wage Councils was reduced. In spite of threats of defiance, court-imposed fines were collected from the NUM, NGA, and Transport and General Workers' Union. Union membership fell from over 13 million in 1979 to 10 million in 1986 (or from 54 per cent to 46 per cent of the work-force) and the main losers were the general unions in manufacturing.

Many analyses of the unions in the 1970s suggested that such political power as they possessed rested on their ability to defy incomes policies over time and veto (as *In Place of Strife*) or render inoperable the policies or legislation which they did not like. The Thatcher governments have gone a long way towards puncturing claims about the power of the unions. They have not sought compromises with the unions over incomes policy or industrial relations; the unions are no longer an 'estate of the realm'. The creation of the legal framework has ended the 'voluntarist' tradition that has dominated industrial relations in the twentieth century. The heavy fines imposed on Sogat 82 in its 1986 dispute with News International showed that the unions' inability to carry out secondary or solidarity 'blacking' weakened their bargaining power. At the same time the government did not push its reforms as far as some supporters wished; it did not, for example, place the onus on union members to contract in rather than out of paying the political levy and has done little to break down national wage bargaining which pays little attention to different local labour market conditions.

Conclusion

A problem with the presentation of the foregoing data is to assess the amount of responsibility that the government has for the policies. After all, privatization, deregulation, tax cuts, falling capital markets, and reducing union power have been found in other states in the 1980s. There have been four different interpretations of the economic record of the Thatcher government to date. The first is that the policies have largely been implemented as intended and that they are bearing fruit. The lower inflation and freer market, it is claimed, has generated an economic recovery. In the years between 1982 and 1988 the economy grew at an average of nearly 3 per cent per year. The large increase in unemployment, though regrettable and unexpected, is largely the responsibility of the trade unions for exacting high wage increases and overmanning in the past. The second interpretation is that the economic outcomes, *unfavourable* as well as favourable, have been a consequence of the government's policies. For example, the massive recession between 1979 and 1981 was caused by too severe a financial squeeze. According to Jackson 'the recession of the 1980s was of the government's own making. Overtight monetary and fiscal policies and the loss of overseas markets, due to an over-valued pound, resulted in an erosion of demand, falling investment, rising inventories, bankruptcies and an explosion in the numbers unemployed.'[20] Some economists calculate that only a small part (perhaps a quarter) of the appreciation of the exchange rate is attributable to oil but that the government deliberately pushed up the exchange rate (which appreciated 40 per cent between 1979 and 1980) by raising domestic interest rates and thereby weakening industrial competitiveness.

A third interpretation is to say that the strategy has not been implemented and therefore the government is not fully responsible for the outcome. Control of money supply (M3) has been erratic, and it has proved difficult to establish a relationship between the changes in M3 and the fall in inflation. A final line of analysis insists that the government has made little difference, particularly on unemployment. James Alt argues that the increase in supply of North Sea oil provided a major boost for sterling which in turn resulted in a loss of competitiveness for British goods and a consequent loss of manufacturing goods and employment. Hence the major cause of unemployment lies outside the government's responsibility and a Labour government would not have made much difference.[21]

The balance sheet for managing the economy is, not surprisingly, mixed. Over the years 1982 to 1988, economic growth has averaged a respectable 2.5 per cent, close to the figure for the 1960s. But this followed the sharp decline in 1980 and 1981; over the period 1979–85 the annual average growth rate has been 1.9 per cent about the same as most other OECD countries—without the benefit of North Sea oil. There have been some impressive cases of increases in productivity, notably British Airways, British Leyland, and British Steel, which have shed workers, but it remains to be seen whether the improvement will be durable or a once-only case of firms shaking out the least efficient workers. Business profits have increased from 3 per cent in 1979 to 10 per cent in 1987. Unemployment over the decade has been a disaster though the total has steadily fallen in the last two years. The much praised reduction on inflation only brought Britain back into line with the OECD average and in 1989 was again above it. By May 1989 inflation at over 8 per cent was at the same level as at the 1979 general election and unemployment, though falling, was higher than in 1979.

The weakening of trade union power in the 1980s has been due to the interaction of government legislation, the rise in unemployment, and technological change. It is too soon yet to make a judgement about the relative importance of these different factors. The government's legislation on industrial relations and trade unions has taken hold. Here is an area where the balance of advantage has changed since 1979, from union leaders to employers and managers and from the consultative role granted to the unions to one in which they are virtually ignored by the government. The reduction in restrictive practices and number of strikes have been other gains. Rising unemployment was important in producing the reductions in unions' members, funds, and militancy. But the unions' unpopularity in the wake of the Winter of Discontent and Labour's electoral decline also weakened the movement. The legislation has made the creation of new closed shops more difficult, has restricted definition of what are legal strikes, and has exposed union funds to claims against unlawful industrial action. The growth of part-time, female, and self-employment, as well as the growth of the service sector has created groups which the unions have long found difficult to organize. Privatization is another area in which the government has been more radical than was anticipated. But there is a limit to the number of profitable assets which can be sold. Revelations in February 1986 that the government was prepared to sell off British Leyland and Austin-Rover to American-owned companies provoked opposition from Conservative MPs in the affected constituencies which killed the proposals. Opinion surveys suggest that both the trade union reforms and the early privatizations were popular with the voters. The support for the status quo in these two areas had been 'soft', with the public acquiescing because there appeared to be no politically workable alternative. The planned private sales of the water and electricity industries in 1990, however, are unpopular with most voters and worry some Conservative back-benchers.

Some measures have 'freed' the economy for the operation of market forces. They include the income tax cuts, union reforms, and ending exchange controls. On the other hand Mrs Thatcher has jealously maintained tax reliefs for home-ownership and private pensions (indeed, extended them for private health care for the over-60s) and privatization of gas and telephones have done little to increase competition.

The government has made minor cuts on welfare and some changes of a largely symbolic nature. It has protected 'real' levels of spending on the NHS and welfare state, and attempts to contain social security expenditure were thwarted until 1986 by the massive rise in unemployment. Although the government has encouraged privatization at the edges of the health service and social security, the opportunity (presented by Mr Fowler's social security review and then the NHS review) to make a fundamental reform of the legacy of Beveridge was passed by. Ministers have been worried about reactions among voters and among Conservative MPs and have so far shown little interest in pushing the more radical free-market policies advocated by the Adam Smith Institute and IEA.

By twentieth-century standards the Thatcher governments have been radical and successful—not only in winning elections but in achieving so much of what they intended. In 1989 the government was still pressing on with further measures of privatization, reforming the legal profession and health service, resisting moves to greater integration within the European Community, and implementing the new system of local government finance and a reformed education system. The experience is one to suggest an affirmative response to the question, 'Do Parties Make a Difference?'

Notes

1. See A. Lijphart, *Democracies* (New Haven, Yale University Press, 1984) and A. King, 'What Do Elections Decide?', in D. Butler *et al.* (eds.), *Democracy at the Polls* (Washington DC, AEI, 1981).
2. For a review of findings, see D. Kavanagh, *Political Science and Political Behaviour* (London, Allen & Unwin, 1982) chap. 8.
3. *Adversary Politics and Electoral Reform* (London, Wigram, 1975).
4. *Do Parties Make a Difference?* (London, Macmillan, 2nd edn., 1984).
5. Ibid., chap. 7.
6. For such reviews see P. Riddell, *The Thatcher Government* (Oxford, Martin Robertson, 1983), M. Holmes, *The Thatcher Government 1979–1983* (London, Wheatsheaf, 1985), and P. Jackson (ed.), *Implementing Government Policy Initiatives: The Thatcher Administration 1979–1983* (London, RIPA, 1985).
7. For details of the report, see *The Economist*, 18 Sept. 1982.
8. See D. Heald and D. Steel, 'Privatizing Public Enterprise', *Political Quarterly* (1982), J. Kay, C. Mayer, and D. Thompson (eds.), *Privatisation and Regulation—the UK Experience* (Oxford, Clarendon Press, 1986).
9. R. Pryke, *The Nationalized Industries* (Oxford, Martin Robertson, 1981).

10. Private interview.
11. M. Pirie, *Privatization* (London, Adam Smith Institute, 1985).
12. See J. Vickers and G. Yarrow, *Privatization—an Economic Analysis* (Cambridge, Mass., MIT Press, 1988).
13. See P. Dunleavy and R. Rhodes, in H. Druker *et al* (eds.), *Developments in British Politics 2* (London, Macmillan, 1988), p. 140.
14. Walters has presented his account of his experiences in *Britain's Economic Renaissance: Margaret Thatcher's Economic Reforms, 1979–1984* (Oxford University Press, 1986).
15. Report I (Cmnd. 6171).
16. See *Social Trends 1988* (London, HMSO, 1989).
17. P. Johnson and G. Starle, *Taxation and Social Security 1979–1989: The Impact on Household Incomes* (London, Institute for Fiscal Studies, 1989).
18. J. Alt, 'It May Be a Good Way to Run an Oil Company, but . . .', unpublished paper presented at Harvard University, USA, April 1985.
19. See the careful discussion in Samuel Brittain, 'The Government's Economic Policy', in D. Kavanagh and A. Seldon (eds.), *The Thatcher Effect* (Oxford, Clarendon Press, 1989).
20. P. Jackson, 'Perspectives on Practical Monetarism', in *Implementing Government Policy Initiatives*, p. 65.

9

MARGARET THATCHER: A MOBILIZING PRIME MINISTER

Mrs Thatcher was the dominant figure in British politics in the 1980s. She has already established a number of landmarks in political history and psephology: the first woman Prime Minister in any major western industrial state; the first leader since Lord Liverpool in the 1820s to win three elections in a row, and Prime Minister for the longest uninterrupted spell in the twentieth century. Alone of twentieth-century Prime Ministers her name has been given to a political 'doctrine' and undoubted political style. She was the first British Prime Minister to mount a sustained challenge to post-war consensus politics and, more than any other post-war Prime Minister, she is often regarded as a figure apart from her Cabinet. The attention that she has received (particularly the hype surrounding her tenth anniversary in office in May 1989) can tempt one to react and downplay the importance, if not the existence, of Thatcherism. Yet her role in altering the political balance in the Conservative party and directing the government has been so important that she merits discussion in her own right.

Mrs Thatcher provides a mobilizing style of political leadership. In some respects it resembles the wartime leadership style of Lloyd George (1916–22) and Winston Churchill (1940–5). This bold personal style distinguishes her, as it distinguished Lloyd George and Churchill, from other Prime Ministers. But unlike them she did not come to office during a wartime emergency nor head a coalition government. The political style of British political leaders has varied, broadly, between the mobilizing and the reconciling. The mobilizer emphasizes taking decisions, task performance, and changing the status quo, whereas the reconciler is more concerned to maintain the consensus and cohesion of a group. The former is mainly concerned with the achievement of goals, not overly concerned about opposition and the costs of disturbance; the latter is more concerned to represent and respond to diverse interests and is willing to arrive at compromises, if necessary sacrificing policy goals.[1]

The earlier chapters of this book have shown how events and a growing sense of disillusion with many old policies paved the way for a new approach. But influential politicians were still required to push the new policies. Mrs Thatcher has been such a figure and, in spite of reverses, she has animated and directed governments with a sense of political direction. Chapters 1 and 4 examined, respectively, her political beliefs and how she came to the leadership of the Conservative party. This chapter analyses her record as Prime Minister; in particular it examines her political style and role in relation to the Cabinet, Parliament, the Conservative party, and the public.

British Political Leadership

Britain is widely regarded as having a political system which scores high on political institutionalization and low on personal leadership. Disciplined political parties and a system of parliamentary and Cabinet government have been conducive to a restrained, impersonal style of political leadership. This style has been reinforced by the values of partisanship and collective leadership of the Cabinet. Unlike France or the United States, the political head of state is not the focus of regime loyalty. Personal leadership is less appropriate for a British Prime Minister who is a servant of the Crown. Lloyd George and Winston Churchill were exceptions, but exceptions that proved the rule. Both were in some respects political outsiders who came to power during a breakdown of 'normal' party politics and exercised a personal leadership at a time of national crisis.

British political science has little or no literature on political leadership. In Britain we refer rather to the office of Prime Minister and his or her performance rather than national leadership or individual leaders. The academic literature on the Prime Minister has given rise to one of the few serious debates in British political science; the idea of Prime Ministerial government versus that of Cabinet government. The first view is associated with the late John Mackintosh and Richard

Crossman and is regarded as being a trend, regardless of incumbent. The post-war period had seen the transformation 'of Cabinet government into Prime Ministerial government'.[2] This view is reinforced by the mass media's tendency to personalize politics and by informal discussions of politics (for example, 'Mrs Thatcher's Government').

A rival interpretation points to the limits operating on any Prime Minister—for example, constraints of time, the greater departmental resources available to other Cabinet Ministers, and the premier's need to keep the Cabinet together, conciliate rivals, and manage the political party in Parliament.[3] The British executive is collective and the Prime Minister's power is exercised *in* and *with* the Cabinet. In addition, there are the inevitable pressures from the outside world, the difficulties of the British economy, and the various lobbies. Limits on British government are limits on the Prime Minister.

Political Style

It is worth restating how *radical* a leader Mrs Thatcher has been for the Conservative party. She was an accidental leader and in some ways did 'hijack' the party. Conservative leaders have traditionally been accommodating to the major interest and sought compromises. After 1951 Winston Churchill and his Conservative successors protected the welfare state, maintained full employment, and conciliated the trade unions. Mrs Thatcher, on the other hand, claimed that previous Tory leaders had agreed on the country's problems and what needed to be done—cutting public expenditure and direct taxation, reforming the unions, and restoring the incentives and financial disciplines of the free market—but failed to carry through the policies. She saw herself as providing the political resolution that had been lacking hitherto. Mr Heath's U-turn in 1972, shifting from free-market policies to a statutory prices and incomes policy and state intervention in industry, is a negative example for her. By the mid-1970s collectivist policies and the constraints on goverment they represented were so deeply entrenched that a virtual counter-revolution was required.[4]

Her image as a confrontational politician stems from her conviction that British politics had become an important, even a decisive, battle of ideas. Mrs Thatcher is one of the few senior politicians who takes pride in stating her political convictions and insists that policies should derive from a coherent set of principles. Without such a starting point, she argues, a leader is at the mercy of events and unlikely to produce coherent policies. When questioned in August 1978 by the present author about the basis of her political strategy, she replied: 'My political beliefs'.

An insight into Mrs Thatcher's character is conveyed by the kind of people with whom she feels comfortable. She admires wealth-creators, entrepreneurs, and people who have taken a chance in the market rather than those who work in the public sector. In her view the provision of resources for the public services depends largely on the efforts of the 'doers' and wealth-creators. Yet the latter, she believes, are too often denigrated in the media and universities. In contrast to the immediate post-war generation of Conservative leaders, Churchill, Eden, Butler, and Macmillan, she has little sense of guilt ('bourgeois guilt' was the phrase she used in New York to the Institute of Economic Studies on 15 September 1975) for the unemployment of the 1930s. But such feelings were important among a number of One Nation Tories in the 1950s. Harold Macmillan, for example, never forgot his experience as an MP in the economically depressed north-east of England in the 1930s. As Prime Minister he accepted the resignation of his Chancellor of the Exchequer in 1958 rather than agree to modest public expenditure cuts. But in the 1970s few self-made first generation Conservative politicians had personal memories of the 1930s, or were any longer inclined to conciliate the trade unions and make the avoidance of unemployment the main goal of economic policy.

In public and private she is a relentless educator. An important task of leadership, in her view, is to win the battle of ideas, and this is done by frequently expressing basic beliefs and principles. She seems to revel in arguments and loses no opportunity to declare her political principles. Initially Shadow Cabinet colleagues were worried by her view of politics as an ideological battle ground and by her disavowal of many cross bench attitudes (Chapter 1). They feared that she would be seen as (indeed *is*) abrasive and lacking compassion. She is famous for berating and lecturing people, ministers, civil servants, TUC delegations, and the House of Commons; 'rubbish' or 'what mean?' and other rude comments are annotated to civil servants' papers. Phrases such as 'TINA' ('there is no alternative') in answer to critics of her economic policy, and 'We want our money' ('I cannot play Sister Bountiful to the Community') to her EEC partners over the perennial problem of Britain's budget payments, sum up the spirit of what she feels and argues.

The warrior image is also aided by her forceful style of speaking and assertive personality. Her self-confidence and certainty make her overbearing in discussion. More than one sceptical senior colleague has been admonished, 'Don't tell me what; tell me how. I know what.' It is as if political goals are what she decides and the role of ministers, civil servants, and advisers is to help implement them. She is impatient at what she regards as generalities in memos from colleagues and civil servants—even when she agrees with them. She wants advice about practical measures which can achieve objectives. Even friends acknowledge that she approaches conversations as an intellectual exchange rather than as an opportunity to empathize; her invitations to colleagues to define their terms and to explain 'precisely' what they mean often disconcert the unprepared. (Former schoolfriends recall an occasion when Mrs Thatcher, as a young MP, returned to her old school as chief speaker at a dinner for Old Girls, and corrected the headmistress, who was a classical scholar, on the pronunciation of her Latin.[5]) David Howell did not remember his time in Cabinet with much pleasure—'some arguments just left such acrimony and ill-feeling that I can't believe they really could have been enjoyable . . . I think the general atmosphere in the government of which I was a member was that everything should start as an argument, continue as an argument and end as an argument.'[6] Her frequent denunciations of high levels of taxation and public expenditure, of big government, and of the diminution of individual freedom and choice are passionate and deeply felt; they are expressed in attacks on the baneful, almost 'immoral', effects of inflation and of governments which debase the currency, or borrow rather than 'balance' their income and expenditure.

For all the pejorative talk of her being an 'ideologue', however, she is a practical Conservative. She traces her beliefs back to her upbringing in Grantham and to the personal experiences of the 1970s, rather than books. 'We tried it then and look where it led us', is her view of previous bouts of big spending, reflation, and subsidies to troubled industries. Colleagues are impressed (and sometimes alarmed) by her willingness to link large philosophical issues with day to day matters; for example, high state expenditure and taxation mean less for the ordinary citizens to spend as they choose. She has boasted of her tendency to draw analogies between running the economy and a housewife's weekly budget (which many economists dismiss as pre-Keynesian, naïve, and dangerous). Some ministers were at first embarrassed at what they regarded as her 'preaching' and 'moralizing' but in the end accepted it as part of the Thatcher style. Unlike Harold Wilson she did not, until the Westland imbroglio in 1986, have the reputation of a 'fixer', compromiser, or manager in her Cabinet. Indeed she once described herself as 'the Cabinet rebel'. Discussions with colleagues and interviewers are often interrupted by rhetorical questions, 'Do you see?' or 'Have I made that clear?' When taxed about her 'bossy' or 'headmistressy' role in the 1983 election she replied: 'Yes, I do believe certain things very strongly. Yes, I do believe in trying to persuade people that the things I believe in are the things they should follow . . . I am far too old to change now.'

Her own radicalism is clearly tilted against certain dominant institutions and interests. On 18 January 1984 she told a gathering of parliamentary lobby journalists that she wished to be remembered for breaking the consensus and tackling traditionally 'immune targets'. These included the trade unions, nationalized industries, local government and, surprisingly for a Conservative leader, the monopoly professions like the solicitors and opticians and, no doubt, the civil service and universities.[7] In the same speech she said that she wanted her government to be remembered as one 'which decisively broke with a debilitating consensus of a paternalistic Government and a dependent people; which rejected the notion that the State is all powerful and the citizen is merely its beneficiary; which shattered the illusion that Government could somehow substitute for individual performance'. Mrs Thatcher is also out of sympathy with the previous Conservative leaders' looking for 'deals' with the major interests, conciliating the trade unions, and concerned to be 'good Europeans' at almost any cost. She has offended previous Conservative leaders like Mr Heath and Lord Stockton (formerly Mr Macmillan). The extent to which her policies and style have antagonized the universities was seen in the decisive vote by Oxford academics not to give the university's most famous female graduate an honorary degree in January 1985. Leaders of the Church of England have also attacked the 'hardness' of the government, only to be dismissed in return as 'cuckoos' (see below, p. 289). The image is of an aggressive radical tilting against many of the most powerful and conservative (even Tory) interests in British society.

Indeed, this is Mrs Thatcher's image of herself. In a lengthy and revealing BBC radio interview on 17 December 1985 with Michael Charlton she defended her brand of Conservatism:

> it is radical because at the time when I took over we needed to be radical. It is populist . . . I would say many of the things I've said strike a chord in the hearts of ordinary people. Why? Because they're British, because their character is independent, because they don't like to be shoved around, because they are prepared to take responsibility . . .

And for the 'trimming' of previous Conservative leaders she borrowed Kipling's words: 'I don't spend a lifetime watching which way the cat jumps. I know really which way I want the cat to go.'

But this determination was coupled with a prudence and caution in deciding which battles to fight. In foreign affairs she is only one national leader among many. She compromised with China in agreeing to the return of Hong Kong to China in 1997 and with the Zimbabwean nationalists over the Rhodesian negotiations, when it became clear that she could not get her way. In June 1989, isolated in the European Community and under pressure from Cabinet ministers, she agreed to set conditions for British entry to the EMS. At home in February 1981 she gave way and provided money for twenty-three uneconomic coal pits to remain open, when faced with the threat of a coal strike. By 1984 the government, however, was prepared for a lengthy coal strike and refused to compromise. In 1982 she disowned a CPRS paper, which suggested drastic cuts in the welfare state including some privatization of the NHS, when the Cabinet reacted negatively. In other words, Mrs Thatcher is prepared to wait for opportunities.

Cabinet Management

The formal duties and responsibilities of the office of a British Prime Minister are not laid down in a constitution or statute. Yet a Prime Minister does face certain persisting role demands—chairing the Cabinet, answering parliamentary questions, making appointments, and choosing the time for a dissolution of Parliament. Although Asquith (Prime Minister 1908–16) exaggerated when he said, 'The office of the Prime Minister is what its holder chooses and is able to make of it', the personality of a Prime Minister may give a distinctive character to his or her administration. This is particularly the case with Mrs Thatcher. Most recent Prime Ministers have usually presented themselves as representing a collective Cabinet viewpoint. Mr Attlee, for example, saw his job as being 'to collect the voices of the Cabinet' and Mr Wilson compared his role as Prime Minister at various times to a soccer 'midfield sweeper', or the 'the conductor of the orchestra'. The cases where Prime Ministers have tried to bypass the Cabinet on a major policy or acted without clear Cabinet support, as for example Neville Chamberlain's foreign policy in 1937–9 or Mr Wilson's attempted trade union reform in 1969, may have reinforced the principle. Government in Britain is not presidential; it is a collective enterprise in which power is shared and for a beleaguered Prime Minister a power shared is a blame shared.

Mrs Thatcher has a clear view of her role as Prime Minister and sees herself as an activist rather than an arbitrator in Cabinet disputes or a spokesman for a collective Cabinet view. There is a great difference between the Attlee concept and Thatcher's robust assertion of 1979: 'It must be a conviction government. As Prime Minister I could not waste time having any internal arguments'[8] or 'Yes, I do drive through things which I believe in passionately—what else do you expect of a

Prime Minister? I'm not here just to be Chairman, I'm here because I believe in things.' (*Panorama*, BBC TV 25 January 1988.) Her job is to push Cabinet ministers 'to do what is right'; this involves reminding them of the Government's strategy laid down in the manifestos and combating what she regards as the inertia inherent in departments. Given the government's commitment to control public expenditure and reduce taxation, her main troubles in the first term were with the major spending departments and with ministers who favoured some further reflation to ease unemployment. In part, this was a consequence of her appointing some ministers who were sceptical of her economic strategy. The most troublesome period concerned the budget in 1981, when Jim Prior, Peter Walker, and Sir Ian Gilmour were so opposed to its deflationary thrust that they considered resigning. Although they neither rebelled nor resigned, their disenchantment was soon widely known. But the target of Mrs Thatcher's wrath was equally well known when the day after the budget in a public speech she said: 'Now what really gets me is . . . that those who are most critical of the extra taxes are those who were most vociferous in demanding extra expenditure.' This was a remarkably public statement of a Prime Minister's resentment of colleagues.[9]

Since then problems with colleagues have centred on particular issues such as Westland (1985–6), exchange rate policy (1988), and British entry to the exchange rate mechanism of the European Monetary System. Mrs Thatcher has insisted on particular lines of policy, even over the doubts of the responsible department. She has strong views on a wide range of issues. Her own policy agenda, as King notes, has often been separate from that of the Cabinet or Conservative party.[10]

British Prime Ministers do not have a department to administer. The opportunity this gives an incumbent to take a broad view is useful if he or she has the resources to know what is going on. The support staff is tiny in comparison with that of a Cabinet Minister or US President—about eighty staff working in Downing Street of whom the great majority (about sixty) are clerical and secretarial staff, messengers, clerks, or administrators of the machine. For all the talk of Prime Ministerial government, one must remember the unequal battle which any incumbent, when confronted by more than a score of powerful departments, faces. The Prime Minister lacks executive powers and has therefore to work with and through ministers who have executive powers vested in them collectively. Number 10 Downing Street, in the words of a former Political Secretary of Mr Heath, is 'a house, not an office'.[11]

Mrs Thatcher's Private Office consists of six fairly senior civil servants who help with official speeches and parliamentary business and liaise between herself and departmental ministers. In addition, there is a Policy Unit of eight or nine political advisers. This body was established by Harold Wilson in 1974 and has been retained by his successors. Its members are political appointees and civil servants who contribute advice on policy and papers from other departments, liaise with the party, and help with speech writing. They are particularly concerned to look for the policy and party political consequences of events and policies. John Hoskyns headed the Policy Unit until 1982. Hoskyns had built up a computer company and been active in the Centre for Policy Studies before joining Mrs Thatcher. He was a vigorous proponent of policies of controlling the money supply, cutting public spending, and reducing the legal privileges of the unions. He retired in 1982 in some frustration with the civil service before eventually becoming director-general of the Institute of Directors. Ferdinand Mount, then the Political Editor of the weekly magazine *Spectator*, replaced him. He was followed in January 1984 by John Redwood, a historian turned merchant banker, who in turn was succeeded by an academic economist Brian Griffiths. In 1970 Mr Heath had established the Central Policy Review Staff to deal with the long-term strategy matters and provide briefs for the Cabinet as a whole. Mrs Thatcher did not have much time for the body and abolished it after the election victory in 1983—an example perhaps of her dislike for so-called 'professional' advice that purports to be above party politics. She had already decided to strengthen her Policy Unit and some CPRS members were recruited to it.

Her reservations about the senior civil service as a whole were well known in advance. In January 1981, worried over failures to control money supply, she recruited Professor Alan Walters from the United States to provide economic advice which was independent of that of the Bank of England and

Treasury. In particular, she felt in need of somebody like Walters who could argue with these two bodies about control of the money supply.[12] Her disenchantment with the Foreign Office, already evident over Rhodesia and the EEC, came to a head during the Falklands War and she brought in Sir Anthony Parsons (who had impressed her during the war when he was ambassador to the United Nations) to advise on foreign affairs and appointed Roger Jackling to advise on defence. This clutch of appointments, together with her post-Falklands dominance in the Cabinet, was widely seen as heralding the introduction of a Prime Minister's Department.

There were predictable objections from departments which did not want to have their own policy advice scrutinized by outside experts. In fact no permanent institutional reform came from the moves, and soon after the 1983 election Parsons and Jackling left, the latter returning to the Ministry of Defence. Professor Walters left in 1985 and returned in 1989. The changes were more a reflection of her anti-establishment outlook and determination to shake up the civil service hierarchy. Shortly after the abolition of the CPRS, Professor George Jones noted: 'While she has been strengthening her own personal staff resources, she has weakened those at the disposal of her Cabinet colleagues for the performances of their collective deliberations by abolishing the CPRS. So she has tipped the system a little away from collective to presidential government'. Jones adds, however, that 'the tipping is only slight'.[13]

But her reliance on non-official advice could also have significant policy consequences. Mrs Thatcher's speech at Bruges in September 1988 which set out the government's view of its relations with the European Community was drafted by her Private Secretary and foreign policy adviser, Charles Powell. Its forceful statement of British sovereignty offended many in the Foreign Office, which is more *communautaire*. Professor Walters has reinforced Mrs Thatcher's opposition to British entry to the European Monetary System (favoured by Sir Geoffrey Howe and Nigel Lawson) and to her Chancellor's policy in 1988 of managing the exchange rate to maintain downward pressure on inflation. In 1988 and 1989 the public divisions between the Prime Minister and her Chancellor were politically and economically damaging.

The view that Britain was moving to Prime Ministerial government was largely encouraged by Mrs Thatcher's dominant personality, which became more marked following the Falklands war. But there were other factors as well: her ruthless dismissal of Cabinet dissenters in 1981 and sacking of the Foreign Secretary in 1983, the abolition of the CPRS, her close involvement in the promotion of Permanent Secretaries and her tendency, by publicly expressing her own views on controversial issues, to pre-empt Cabinet discussions. She has reduced the number of Cabinet and official Cabinet committee meetings, and fewer papers are distributed to the Cabinet—the very stuff of collective decision making.[14] She has relied more on her Policy Unit and *ad hoc* groups, and intervened more in departments. Mrs Thatcher is an activist Prime Minister and her Press Office is keen that she is seen as such. 'MAGGIE ACTS' or 'MAGGIE STEPS IN' are familiar tabloid headlines as Mrs Thatcher convenes seminars of experts at Downing Street to cope with a pressing issue. This has occurred with football violence, broadcasting, 'green' issues, schools, and the universities. Yet this interventionism has not led to the creation of anything approaching an Executive Office. Although there is a general trend for the private offices of many political heads of state to increase in size and influence, the British Prime Minister still appears to be less well endowed with this type of political support than leaders of most other states. It is possible to argue that, compared to the staff available to previous peacetime premiers, today's staff at Number 10 is actually smaller in relation to the greater burden of work.

Government decisions in Britain are made in the name of the Cabinet, in contrast to the United States where they are made in the name of the President. The British Cabinet consists of most of the ruling party's political 'heavyweights', politicians who have either an acknowledged expertise, political following, reputation, or are close to the Prime Minister. Most of the twenty or so men and women in the Cabinet are well known to one another and have worked together over some years. But most are also rivals for political promotion to a higher office and, ultimately, for the Prime Minister's job.

Two points are noteworthy about Mrs Thatcher's early appointments. First, in making her appointments in 1975 to the

Consultative Committee or Shadow Cabinet, Mrs Thatcher displayed conciliatory traits in retaining so many of Mr Heath's appointees. All but three had already served under Mr Heath. Of the twenty original members it is likely that only two (Sir Keith Joseph and Airey Neave) voted for her on the first ballot for the leadership election in 1975. A Shadow Cabinet containing senior figures like Lord Carrington, Whitelaw, Prior, Maudling, and Pym—most of whom had already achieved office under Mr Heath, were associated with his policies, and owed little to her—was hardly right-wing. Although her appointments were made with an eye to preserving party unity as well as recognizing ability, Thatcherites—often young and lacking ministerial experience—were not obvious candidates for promotion at the time.

The second noteworthy feature is her initial caution in refusing to reshuffle spokesmen. Most of the 'Shadows' remained in post for the four years of opposition. Michael Heseltine was moved from Trade and Industry to Environment. But, apart from the dismissal of Reginald Maudling as Foreign Affairs spokesman, the only important reshuffle followed John Davies's resignation on grounds of ill health from Foreign Affairs in November 1978. Francis Pym assumed this post, Norman St John-Stevas replaced Pym as Leader of the House of Commons, and Mark Carlisle in turn took over Education from Mr Stevas. All the spokesmen were appointed to the Cabinet in 1979, usually to departments which they had shadowed in opposition. Mrs Thatcher, like Mr Heath before her, clearly regarded opposition as a time of preparation for government.

However, continuity also meant that Mrs Thatcher was in a minority over some policy battles in the new Cabinet. Some ministers were openly disgruntled with her style of leadership and the Treasury's economic strategy. There was a clear separation between the Cabinet and Mrs Thatcher in the minds of many Conservative activists and observers. Many right-wing back-bench critics of the government (but supporters of Mrs Thatcher) could voice their criticism without including her and liked to present her as a political prisoner of Cabinet 'Wets'. After a while, Mrs Thatcher herself, in coded language and indirect statements (through some of the Number 10 staff), may have provided encouragement for them. But a

Prime Minister's power of appointment is not limited to the selection of ministers; equally important is the allocation to particular posts. Mrs Thatcher was careful on the whole to appoint her supporters to the important economic ministries. Sir Keith Joseph went to Industry, Sir Geoffrey Howe and John Biffen to the Treasury, John Nott to Trade. Such likely opponents of the economic strategy as Lord Carrington and Sir Ian Gilmour were banished to the Foreign Office while Francis Pym went to Defence. She also ensured that she chaired the important Cabinet (later 'EA') Committee on economic strategy and that it contained a majority who supported the economic strategy of herself and the Chancellor.

During the course of 1981 Mrs Thatcher dismissed a number of notable Cabinet dissenters (Stevas, Soames, Gilmour, and Carlisle), Mr Prior was moved to Northern Ireland, and Lord Carrington chose to resign in April 1982 in the wake of the House of Commons' criticism of the Foreign Office policy preceding the Argentine invasion of the Falklands. By autumn 1981 all the major economic ministries (Energy, Employment, Trade and Industry), apart from the Treasury, had new ministers. The reshuffle was a notable display of Prime Ministerial power but perhaps also of her frustration with the original appointments.

In the new government of June 1983 the move to a more homogeneous Cabinet was continued. Francis Pym, a sceptic with regard to economic strategy, was sacked. In 1987 another sceptic, John Biffen, Leader of the House, and one who publicly called for a consolidation rather than extension of existing policies, was dismissed. Earlier he had been described by Downing Street officials as only 'a semi-detached member of the government'. There was promotion for Nigel Lawson, Leon Brittan, Cecil Parkinson, and Norman Tebbit. All were younger middle-class or lower-middle-class meritocrats; they were tough-minded types of Conservatives, who supported Mrs Thatcher's policies, and owed their promotion to her. Nicholas Edwards at the Welsh Office, George Younger at the Scottish Office, and Mrs Thatcher herself had remained in the same posts since May 1979. With Lord Hailsham's retirement as Lord Chancellor in 1988, Mrs Thatcher alone had remained in the same post and only Peter Walker, George Younger, and Sir Geoffrey Howe had been in the Cabinet continuously.

Over time she gradually created a Cabinet whose members owed their promotion to her. In 1979, the Cabinet of twenty-two contained only seven ministers (Edwards, St John-Stevas, Younger, Nott, Biffen, Carlisle, and Maude) who had not been in a previous Cabinet. The leadership strata of the Conservative party in the 1980s has been shaped by Mrs Thatcher. By January 1989 nineteen of her twenty-one Cabinet ministers owed their first preferment to her. By this date, however, the political balance in the Cabinet had altered. The results of the various appointments including the 1985 promotions of Douglas Hurd, Kenneth Baker, and Kenneth Clarke and the 1987 departures of Joseph and Tebbit gave the Cabinet a distinctly non-Thatcher look. This was confirmed by the promotion of Chris Patten and the departures of Lord Young and John Moore in July 1989. Apart from Nigel Lawson, Nicholas Ridley, Cecil Parkinson, and Sir Geoffrey Howe, it was difficult to think of other Cabinet ministers who would be regarded as economically 'dry' or ideologically 'Thatcherite', and both Lawson and Sir Geoffrey had well-publicized differences of opinion with Mrs Thatcher.

Mrs Thatcher's style certainly contributed to the Cabinet divisions between 1979 and 1981 and again after 1987. She has strong views on most issues and her propensity to express her views boldly at the outset of a Cabinet discussion, combined with a sometimes dismissive attitude towards opposing colleagues, tended to change the atmosphere and polarize Cabinet discussions. This contrasted with the approach of some other Prime Ministers. Attlee, Macmillan, Wilson, and Callaghan often waited for a policy line in Cabinet discussion to emerge before committing themselves. They also regarded their influence as a finite resource to be husbanded, used on major issues, and not frittered away on minor skirmishes or issues on which they were likely to be in a minority in Cabinet.

Mr Attlee was careful to position himself with the majority view in Cabinet. Mr Macmillan was, according to colleagues, prepared to give way with good grace when he could not carry the Cabinet. He intervened on 'big' issues or government initiatives (such as applying for membership of the EEC, and incomes policy, defence, and foreign and Commonwealth matters). Mr Heath was a directive leader of the Cabinet but confined his interest to a few areas (EEC, incomes policy, and Northern Ireland). By contrast, Mrs Thatcher leads from the front, has views on most issues, and is quick to voice them. After 1987 she publicly disagreed with policies of her Foreign Secretary and Chancellor.

For much of her first term the thrust of economic policy was regarded sceptically by many colleagues. Monetarism was a particularly controversial policy and a number of colleagues argued for a reversal of strategy, as unemployment increased remorselessly and the government's popularity in opinion polls and by-elections slumped. For some months in 1981 the Chancellor, Sir Geoffrey Howe, had lost the confidence of Cabinet colleagues. Mrs Thatcher's public speeches contained thinly veiled warning messages to colleagues who doubted the strategy. Statements such as 'The Lady's not for turning' and 'Are we going to go back? Hell, no, we've only just got here' were directed at them. In the first two years of office she had to give way to Cabinet pressure on pay rises for MPs, the scale of public spending cuts in November 1981, gas prices, the Rhodesian settlement, the compromise on the EEC budget (faced with the implicit resignation of Lord Carrington, who had negotiated it, if she did not accept) and, while Mr Prior was at Employment, action against trade union immunities and the closed shop. She also had to bow to pressure from colleagues and permit Cabinet discussion of economic strategy in 1982 and to abandon Cabinet discussion in September 1982 of the paper from the Central Policy Review staff about the implications of the rising trend of public expenditure and its proposals for curbing welfare spending (see above p. 213). It was a remarkable catalogue of rebuffs. Her experience was not one which supported theories of the power of the Prime Minister. She was also impatient with a number of departments for displaying less than whole-hearted devotion to her policies. These included the Foreign Office (allegedly too pro-EEC and generally too prone to make diplomatic concessions at the cost of British interests), the Civil Service Department (too accommodating during the civil servants' pay strike in 1981), the Department of Employment (allegedly too 'soft' under Mr Prior), and Education (allegedly too sympathetic to the education lobbies). Over time she appointed more congenial ministers to run departments. Her position, and that of the Treasury, was fortified by election successes and economic recovery. But she has remained suspicious of departmental lines which resist changes, and her doubts about the Foreign Office have remained.

Whitehall departments have their own networks of interests, hierarchies of civil servants, attentive public and parliamentary groups, and specialist teams of mass media reporters. As a consequence British Prime Ministers may have to work extremely hard to get them to change course. They may intervene in the work of some departments but only at the cost of neglecting broad strategy and the work of others. At the end of the day most departments have to be left alone to implement policies. What is remarkable is that Mrs Thatcher has interfered in departmental matters to a greater degree than many of her predecessors. She has busied herself in the promotion of senior civil servants (looking for people with energy and commitment) and regularly badgered departments about progress on particular policies—'like a dog after a bone' an adviser claims. She suspects that ministers are too often 'captured' by their civil servants. Having won the Cabinet's acceptance of the economic strategy, she has been more concerned with implementation and this appears to have given a greater role to her Policy Unit, members of 'think tanks', and *ad hoc* groups of officials, advisers, and ministers.

It is interesting to study how Mrs Thatcher eventually came to dominate the Cabinet, bearing in mind that many Cabinet ministers in 1979 were not supporters of her economic strategy. An examination of some of the reasons may provide an insight into the sources of a modern Prime Minister's power.

One tactic she has used is to decide matters outside the formal Cabinet, either in committees or in informal groups. She also exploited her control of the agenda of Cabinet and powers to appoint members of Cabinet committees or groups and draw up their terms of reference. The appointment of supporters on the Cabinet committee dealing with economic strategy has already been noted. She chairs two of the most important *ad hoc* Cabinet committees, OD, dealing with foreign affairs and Northern Ireland, and EA, covering economic strategy, privatization, trade union legislation, and the European Community matters. The reduction in the number of Cabinet meetings (45–50 per annum, about half of the post-war norm), and of Cabinet papers (60–70, or one-sixth of the figure in the 1950s), and the appointment of fewer Cabinet committees have reduced the opportunities for collective deliberation. The purchase of the Trident Missiles programme by the government, for example, appears to have been effectively decided by a small group of Cabinet ministers, outside the Cabinet, in early 1981. This involved a massive commitment of future financial resources, as well as deciding British defence policy for the rest of the century. Once the Prime Minister and a few senior ministers had made up their minds, it then achieved what senior civil servants call 'momentum'. The presentation of annual budgets is left to the Chancellor of the Exchequer, though he is careful to consult with the Prime Minister about the strategy and discuss changes beforehand with ministers whose departments are likely to be affected. The final contents of the budget are only revealed to the Cabinet the day before the Chancellor presents it to the House of Commons, when it is too late for any major changes. But two budgets—1979, which included a doubling of the standard rate of VAT on many goods and services and 1981, which was severely deflationary at a time when over two million were out of work—shocked a number of Cabinet ministers. They complained about being 'bounced' into accepting last-minute decisions, were widely reported in the media, and they forced Mrs Thatcher to agree to a Cabinet debate about the economic strategy. However, they had little effect on changing it.

Finally, the decision to ban trade union membership for workers at the intelligence gathering centre (Government Communications Headquarters) at Cheltenham in 1984 was taken outside the Cabinet and Cabinet committees. The decision was initially taken by Mrs Thatcher, Sir Geoffrey Howe (Foreign Secretary), and Michael Heseltine (Minister of Defence). A few other ministers were involved at a later stage. Yet, according to *The Times* (7 February 1984), 'The first most Cabinet ministers knew was when Sir Geoffrey Howe announced the decision in the Commons on January 25.' Many later policies—the poll tax, abolition of the GLC and metropolitan counties, and the health service review have all had a similar Prime Ministerial push. She is not unique in this. It is the scale of her interest in the work of other departments and working with *ad hoc* groups, that makes her distinctive.

Mrs Thatcher's position was also secure because the Cabinet 'Wets' were never a coherent group; the term mistakenly unified diverse critics. Mr Prior, Mr Walker, and Sir Ian Gilmour opposed many of Mrs Thatcher's and the Chancellor's economic policies, and all three at some time seriously considered resignation over the 1981 budget. But they differed over the emphasis they accorded to incomes policies, reflation, or tougher measures against the trade unions. Willie (later Lord) Whitelaw was a loyal deputy. He worked for compromises between the 'Wets' and Mrs Thatcher, but at the end always backed Mrs Thatcher. His support was crucial. The ministers who were dismissed seemed unlikely to lead backbench revolts. But Peter Walker and Jim Prior, also well-known doubters, were retained (Prior until September 1984), no doubt for fear that they might prove troublesome on the back bench. Outside the Cabinet the rejected Mr Heath became an even more isolated political figure. It is very difficult today to imagine a Prime Minister in good health, being deposed by Cabinet colleagues. The effort is likely to be politically disastrous, even if they can agree on a successor.

The above are all classic instruments by which a Prime Minister can outmanœuvre the Cabinet. The first two (deciding Cabinet membership and the members and agendas of committees) lie within the Prime Minister's control, while the third (events) and fourth (divisions among party dissenters) worked to her advantage. There was also the undoubted increase in her personal self-confidence and stature as a result of the recapture of the Falklands and then the 1983 and 1987 election victories.

Mrs Thatcher's first government was plagued by leaks—a symptom of divisions and poor morale. In so far as leaks advertise unhappiness about a line of policy they undermine the principle of collective responsibility, as well as the confidentiality of proceedings. Although she has deplored leaks —and instituted more leak inquiries and prosecuted more civil servants for breaches of confidentiality than any of her predecessors her press office appears to have developed the technique of leaking against fellow Cabinet ministers to a fine art. Mr Prior frequently had reason to complain about hostile leaks from 10 Downing Street when he was in charge of Employment and then of Northern Ireland. Francis Pym's dismissal from the office of Foreign Secretary was widely reported in advance during the 1983 general election; the dismissals of Patrick Jenkin at Environment and Peter Rees at the Treasury in September 1985 were extensively trailed over the summer; the same was true of Mr Biffen's dismissal in June 1987. In the Westland affair her management of Cabinet was an issue, once it emerged that staff at Number 10 had connived with Leon Brittan's office to leak selections from a confidential letter from the Solicitor General to damage Mr Heseltine. The high profile of her press officer Bernard Ingham and his disparaging remarks about Cabinet dissenters or ministers who are due to be sacked have angered some Conservatives. Mr Biffen in 1989 compared the office to a 'sewer' and Mr Pym in 1983 spoke of 'mischief-making by the Prime Minister's poisonous acolytes'. After the Cabinet reshuffle in July 1989 and her press office's rubbishing of the significance of the title deputy prime minister for Geoffrey Howe, Conservative back-benchers went out of their way to show their support for Sir Geoffrey Howe in the House of Commons. Mrs Thatcher also has taken risks with collective responsibility. Mr Whitelaw, as Home Secretary was furious at her applause for critics of his policies at the annual party conference. Mr Prior, at the Department of Employment, was publicly reprimanded by her, and her dissatisfaction with particular policies is usually well known. This public image has been important in fostering a perception of Mrs Thatcher as somebody apart from the Cabinet.

Mrs Thatcher's direct, even confrontational style of managing the Cabinet became an issue in the circumstances in which Michael Heseltine resigned in January 1986. For a month there had been a damaging public dispute between Mr Heseltine, who backed a European Consortium bid to link with Westland, the British helicopter company, and Mr Brittan, the Trade and Industry minister, who favoured an American bid. In the past, Mrs Thatcher had easily fended off challenges from Tory 'wets'. When, however, internal opposition was determined, as Heseltine's was, the danger materialized. In his resignation press conference Michael Heseltine complained explicitly about Mrs Thatcher's style and performance as Prime Minister, claiming that she had acted unconstitutionally in refusing him permission to discuss the issue in Cabinet and in forbidding him to restate views he had publicly expressed in the past. There is a long list of sacked ministers who have objected to her style. Sir Ian Gilmour mourned 'the downgrading of Cabinet government'. David Howell complained that in Cabinet there was 'too much argument and not enough discussion' and regretted the trend towards 'a huge argument where tremendous battle lines will be drawn up and everyone will be hit on the head'.[15] After the damaging Westland affair and the loss of two senior Cabinet ministers, there were even Conservatives who thought that it might be time to change the leader. In its aftermath Mrs Thatcher apparently took greater care to consult the Cabinet more fully and was overruled by it on the proposed sale of British Leyland to the US-based General Motors. After the 1987 general election victory complaints of high-handedness surfaced again. Mr Biffen spoke of 'Stalinism'. What was remarkable was the public display of her willingness to second-guess her most senior ministers, including the Home Secretary, on measures to combat football violence, the Chancellor on anti-inflation policy, and the Foreign Secretary on a wide range of issues. As noted earlier Mrs Thatcher was virtually vetoing European policies favoured by her two most senior ministers. This is not, à la Harold Wilson, a determination to outmanœuvre the Cabinet almost as an end in itself. Rather it reflects Mrs Thatcher's determination to gain space and capacity to achieve her policy objectives.

Parliament

Parliament is the arena in which British political leaders are recruited and it is difficult to think of a successful Prime Minister who has not been able to command the House of Commons. Compared to political leaders in other states, British Prime Ministers spend a good deal of time in the legislature. The lengthy period which most party leaders have already spent in Parliament before gaining office certainly gives them time and opportunity to acquire suitable skills. The parliamentary set pieces like Prime Minister's Question Time (each Tuesday and Thursday when the House of Commons is sitting) and the major debates are great tests for the Prime Minister. At Question Time the Prime Minister is backed by the civil servants who brief her and try to anticipate supplementary questions. Mrs Thatcher takes these sessions seriously and goes to enormous lengths to prepare for them. The noisy exchanges in the House of Commons and her determination not to be shouted down by the opposition make her sound hectoring and strident. John Cole, the Political Editor of the BBC, commented in February 1985: 'To see the Prime Minister, arms akimbo or leaning far across the dispatch box to bellow into the microphone, is to recall a Belfast working-class politician . . . who boasted that his Ma could beat any woman in the street. This lady's not for shouting down.'

In the post-war period, foreign affairs and economic issues have tended to dominate the Prime Ministers' parliamentary attention. Analysis of Prime Ministers' involvement in parliamentary debates shows that in the course of an average year they will participate in six debates and make six statements on policy, usually on the economy, foreign affairs, and government business. Compared to earlier Prime Ministers, Mrs Thatcher makes fewer interventions in House of Commons debates and makes fewer set-piece speeches.[16] Involvement in these issue areas, together with the need to attend to management of the political party, means that she normally sees the Chief Whip, Foreign Secretary, and Chancellor of the Exchequer on political business more frequently than she does other Cabinet colleagues.

The Prime Minister is also the leader of a political party in Parliament. In the British system, leadership of a major political party is the first requirement for being a Prime Minister. A politician, however eminent or popular, who lacks that base will not reach or survive at the top. A party leader who hopes to reach 10 Downing Street has to devote much time to party management, and policies and appointments have to be made with an eye on the reactions of party factions. Of Conservative leaders in the twentieth century, A. J. Balfour (1911), Austen Chamberlain (1922), Neville Chamberlain (1940), Sir Alec

Douglas-Home (1965), and Mr Heath (1975) were all eventually forced out of the leadership because of the lack of party support in Parliament; the last two as much because they had also lost general elections.

Party management in the Commons is left to the Whips, who rely on a mixture of patronage, appeals to party loyalty, and back-benchers' fears of letting the opposition into office to keep the party's majority intact. For the Prime Minister and the Cabinet, carrying Parliament means, effectively, carrying their party's back-benchers. Yet, her first government (1979–83) faced significant back-bench revolts on the proposed closed-shop amendment in the Employment Bill in 1980 and on unemployment benefit cuts in 1981, was defeated on proposed changes in immigration rules in 1981, and had to withdraw a proposal to impose a ceiling on local rates in the face of opposition. Dissent increased in the 1983 Parliament, partly because of the government's larger majority, which enables rebels to dissent without jeopardizing the government. Mrs Thatcher was rebuffed, again largely by her own back-benchers, over the choice of Speaker in 1983, MPs' pay, proposed housing benefit changes, rate support grant proposals, Britain's EEC contribution, and, most humiliating of all, over Sir Keith Joseph's plans to make major cuts in aid to students in higher education. In July 1985 the government was nearly defeated in the Commons over a vote to reduce the Lord Chancellor's salary. This was a Labour vote of censure on the government's acceptance of the proposals from the Top Salaries Review Board for large increases in the pay of senior public servants. In April 1986 the government lost its Shops Bill, which had been designed to permit trading on Sunday. A remarkable example of back-bench influence on the Prime Minister was seen in the resignation of Leon Brittan from Trade and Industry in the aftermath of Westland. Mrs Thatcher had rebuffed suggestions that he resign and made it clear that he enjoyed her full confidence. But it was vigorous criticism of him in the 1922 Committee by Tory back-benchers which convinced Brittan—and Mrs Thatcher—that he no longer enjoyed the confidence of the party in Parliament. This was a hard period for the Prime Minister—she lost the confidence of senior colleagues, was widely regarded as an electoral liability and lost battles in the Cabinet. Within the first year of the 1987 Parliament a quarter of Conservative back-benchers rebelled against the government at least once. There have been major rebellions against the imposition of charges for eye and dental tests, for an amendment to grade the poll tax according to ability to pay, and for the second reading of a private members' bill to liberalize the Official Secrets Act.

The divisions which accompanied Mrs Thatcher's election to the leadership in 1975 have, with one notable exception, faded. Mr Heath, though lacking political influence, has remained a powerful defender of his own government's record —which Mrs Thatcher and some colleagues scorned. His criticisms have become so predictable and strongly worded that they are counter-productive. Mr Heath returned to the limelight in the Euro-election in June 1989. He was bitterly critical of Mrs Thatcher's authoritarian leadership as well as her 'negative' attitude to the Community. His views had some support in the party. Over time, the wets who at one time favoured a change in economic policy virtually disappeared. An attempt by Francis Pym and some thirty dissatisfied MPs in 1985 to form a Centre Forward group to press for such changes was short-lived. The large number of Cabinet ministers who have resigned or been dismissed have not been a focus for dissent. Mr Heseltine has been careful to support the government, while proclaiming a distinctive brand of Conservatism. Another dismissed Cabinet minister, Mr Biffen is a frequent critic of Mrs Thatcher's tone and of the government's preference for tax cuts over greater spending on the social services. But there is little organized dissent.

Mrs Thatcher has always realized that her power base lay with back-benchers and has tried to be accessible to them. In opposition she particularly urged her front-bench spokesmen to be attentive to the party's back-bench committees relevant to their subjects. In part the style is an acknowledgement of how she came to the leadership, and in part also a reaction to previous leaders. Churchill, Eden, Macmillan, and Home had been political grandees; all emerged through the 'magic circle' and not as a result of election by MPs. Mr Heath was notorious for his remoteness from his back-benchers.

She also finds strong support among the party's grass roots.

Like many local constituency Conservative party officers, who are drawn from the professions and small business, Mrs Thatcher is self-made and does not come from one of the traditional Tory interests—business, land, or a political family. Her open dislike of direct taxation, public spending, much of the public sector, and trade unions and her protective attitude to the tax reliefs for home-owners and investors and sympathy for farmers, small business, and the forces of law and order are all favourite Tory themes. She has also proved more willing to allocate political honours to party workers than was Mr Heath who had been particularly niggardly in this regard.

Her invocation of themes and support for many policies favoured by the party's grass-roots supporters has made the party more radical. She has given voice to those who thought that previous Conservative governments had been too tolerant or 'defeatist' about the unions, the welfare state, high taxation, and public spending, and too accommodating to other countries, particularly in the European Community and the Commonwealth. This has been coupled with policies for a 'strong' state in defence and law and order, and more centralized control over local government spending and over secondary and higher education than any previous government has attempted. She is unsympathetic to 'liberal' measures to promote more 'open government', greater press freedom, or trade union rights for government employees engaged in defence and security activities. In this sense she is not a libertarian. The spirit of Mrs Thatcher and the new Conservatism has been well captured by a Marxist, Andrew Gamble:

> The state is to be rolled back in some areas and rolled forward in others . . . The real innovation of Thatcherism is the way it has linked traditional Conservative concern with the basis of authority in social institutions and the importance of internal order and external security, with a new emphasis upon free market exchange and the principles of the market order.[17]

In foreign affairs Mrs Thatcher has proved to be something of a 'little Englander'. In her early years as Prime Minister she largely left the negotiations over Rhodesia (Zimbabwe) and the EEC budget to the Foreign Office. Her deep involvement in the Falklands war and determination to retain British sovereignty over the islands has to be set against Britain's negotiated return of Hong Kong to China with effect from 1997 and refusal to support or condemn the US invasion of Grenada (part of the Commonwealth). Indeed there is some evidence that only with the Argentine invasion of the Falklands and the deaths of British servicemen did the 'Fortress Falklands' policy emerge. Britain has proved to be one of the least 'community-minded' members of the EEC and has for long been at odds with other states about the level of British contributions to the EEC budget. Mrs Thatcher resisted attempts by Brussels to promote a Community-wide Social Charter, which she regarded as introducing 'socialism by the back door'. She also resisted the wishes of senior ministers for Britain to set a date for entry to the EMS. Her strong views on national sovereignty and limiting the competence of the European Community—at a time when moves to greater unity were gathering pace—placed Britain in a familiar minority of one. Yet she did compromise over the budget at Fontainebleau in 1984, signed the Single European Act in 1986, and accepted the Delors proposals for budgeting and agriculture reform in 1988. On Europe the pattern has been one of resistance to proposals but, in the end, accepting compromise.

Although enjoying a warm personal relationship with Mr Gorbachev, the American and West German wishes to promote disarmament on NATO short range missiles in 1989 have left her relatively isolated in the changed climate. Her greater sympathy for the Atlantic connection over the European one is not widely shared by colleagues. In 1989 she described the relationship as 'paramount'. The close relationship was partly a consequence of her rapport with President Reagan, and is unlikely to be as close with President Bush.[18]

Public Role and Image

British Prime Ministers are not popular figures (Table 9.1). They are seen in a partisan perspective, a perception that limits the extent of their popular appeal. Only for five years between 1945 and 1983 has approval for the Prime Minister exceeded that of his party by more than 10 per cent, and Harold Wilson, and Harold Macmillan are the only post-war premiers to have

TABLE 9.1 *Approval for Prime Ministers 1955–1987 Gallup Poll* (in percentages)

Parliament	Prime Minister (Average)	High	Low
1955–1959 (Eden, Macmillan)	55	73	30
1959–1964 (Macmillan, Home)	51	79	35
1964–1966 (Wilson)	59	66	48
1966–1970 (Wilson)	41	69	27
1970–1974 (Heath)	46	45	31
1974–1979 (Wilson, Callaghan)	46	59	33
1979 (June)–1987 (June) (Mrs Thatcher)	39	53	28

Note: All figures are the average (mean) monthly rating.

Question: 'Are you satisfied or dissatisfied with—as Prime Minister?'
Source: Gallup.

retained the support of 50 per cent or more of the electorate for two successive years. Mrs Thatcher's approval rating on Gallup fell to a record low of 28 per cent in October 1981. Although the Falklands improved her standing, the approval ratings over her premiership are still lower than those for leaders in the 1950s and early 1960s. Pre-Falklands, her average monthly approval score on Gallup was 36 per cent; after the war until June 1983 it was 47 per cent. Thereafter it declined to pre-Falklands levels.

Political leaders, where they can be set apart from the party, may help to shape a party's image. In Britain, however, they have a limited impact on voting behaviour. In 1979 Labour decisively lost the election, although Mr Callaghan was more popular than Mrs Thatcher (and his party). The traditional forces for voting, namely party loyalty and social class, though weakening, still depress the electoral impact of the personality of the leader.

Yet mass media coverage of general elections and parliamentary politics is highly personalized and concentrates on the party leaders. General elections have become more presidential and the mass media virtually ignore the secondary party leaders.[19] For the past quarter-century, the Prime Minister alone has regularly received more attention in *The Times* than the three leading ministers together—the Foreign Secretary, Chancellor of the Exchequer, and Leader of the House of Commons.[20] Television and the popular press are even more likely to accentuate the pattern of *The Times* reporting. The imbalance in coverage stems partly from the Prime Minister's engagement in more newsworthy activities and partly from the fact that, as leader of the government, he or she is newsworthy. He has many opportunities to personify the government or, strictly speaking, to be presented as its spokesman. In the mass media, on foreign visits, and in Parliament, the Prime Minister is regarded as the authoritative spokesman for Cabinet policy. Political coverage on television necessarily promotes leaders because it usually has time for only one item per party. Mrs Thatcher was particularly prone, as opposition leader and early in her premiership, to 'make policy' in television and other interviews, almost committing her Cabinet in advance. This has continued as Prime Minister. Aided by her Press Officer, Bernard Ingham, the media more and more present government policies as Mrs Thatcher's. Ministers have complained about 'the Number 10 publicity machine'.

Surveys indicate that Mrs Thatcher is not regarded as a warm or compassionate person. But she has long scored highly on the qualities of being decisive, resolute, and principled, a perception helped by the Falklands war and the comparison until 1983 with Michael Foot, the Labour leader. The Conservative election campaign slogan of 'The Resolute Approach' in 1983 leaned heavily on her image.

Not surprisingly, Mrs Thatcher has polarized public opinion more than other party leaders. People are not indifferent; they either like or dislike her. In the 1983 and 1987 elections many Conservative and Labour politicians reported the doorstep hostility to Mrs Thatcher, as well as the respect for her. Her own perception of herself (courageous, persistent, and sticking to a course no matter how unpopular) is matched by the public perception, according to Gallup surveys on the personality images of various leading British politicians. Table 9.2 shows that for nearly ten years now voters have agreed about her strength of personality and outspokenness. But many also see

TABLE 9.2 *Mrs Thatcher's Image* (in percentage of people asked)

	Oct. 1977	Apr. 1979	May 1983	July 1987	Positive/ negative shifts
She is a strong personality	84	79	94	96	+12
She speaks her mind	83	81	90	90	+7
She's not in touch with the working-class/ordinary people	54	52	63	70	−16
She divides the country	31	37	60	68	−37
She is a snob, talks down to people	40	37	52	56	−16
Her ideas are destructive, not constructive	23	25	42	41	−18

Source: Gallup.

her as divisive, condescending, and a destructive force in politics. The image of her as a 'superwoman' is also fuelled by eulogics in the sympathetic tabloid press, as well as the 'devil theory' of Labour politicians.

It is a tribute to the importance of her personality that her advisers have spent much time on her image. They were originally concerned that she appeared too aggressive, shrill, and unsympathetic. Before the 1979 election her public relations adviser, Gordon Reece, coached her to lower her voice, speak slowly, and present a 'soft' image and tried to arrange television interviews with relatively non-aggressive interviewers. On the whole she has been more attentive to the requirements of the arts of self-presentation and the requirements of the mass media and public relations than other leaders like Heath, Callaghan, and Foot. A recent study has documented the many ways in which Conservative spokesmen have tried to manage the media and been conscious of the image of the government and Mrs Thatcher. The study concludes that 'She has become a mistress of the pre-planned, carefully packaged appearance.'[21]

The British Prime Minister is inevitably pushed to play a major public role—as head of the government in Parliament, as public spokesman for the government, as leader of the nation on visits abroad, and as chairman and summarizer of Cabinet discussions. But compared, say, to a President of the United

States or France, the tasks of cultivating and presenting the Prime Minister's personality are probably less demanding. On the other hand, tasks of party management and maintaining Cabinet cohesion are more insistent, though perhaps less so than for a German Chancellor or Italian Premier.

Conclusion

Mrs Thatcher has already proved herself a remarkable figure. She has challenged certain long-established political assumptions: that a government could not be re-elected if it presided over a massive increase in unemployment; or that a government had to govern with the consent and co-operation of the major interests, particularly business and the unions. Mrs Thatcher has successfully defied both beliefs. Most British Prime Ministers and party leaders have gradually lost popularity over time and retired when their reputation was at a low point. After a decade in office Mrs Thatcher was still an exception to this trend and was as popular as when she first became Prime Minister.

It is true that she has not always got her own way in Cabinet, particularly in the first years. On a number of issues informed observers were able to distinguish her own preferences from those of most of her Cabinet, hardly a sign of successful political management. Her remarkable statement in an interview, 'I am the Cabinet rebel' suggests that, if she gave a lead, she did not always have a majority of followers. However the fact that she lost a few battles in Cabinet does not mean that she was not the dominant figure or that the Cabinet was supreme.

Reference was made earlier to the *mobilizing and reconciling* styles of leadership. Effective political leaders have to take account of both role demands. It is possible, however, to classify most British political leaders according to the relative importance they attach to these values. In the inter-war years Stanley Baldwin and Ramsay MacDonald, by temperament and political conviction, were reconcilers. They were relatively non-partisan and pursued policies which would least disturb the social fabric. Gladstone. Lloyd George, and Chamberlain were classic cases of mobilizers. They were not respecters of party lines and the last two in particular were widely distrusted among parliamentary colleagues. The mobilizing style has a greater appeal when there is dissatisfaction with the status quo and a feeling that new policies are required.

There is little doubt that Mrs Thatcher belongs to the camp of the mobilizers. She is impatient with the status quo and the traditional style of decision-making by compromising with the major interests. She has repudiated policies associated with previous Conservative leaders and has divided public opinion. She has been highly critical of many established institutions, such as the senior civil service, Foreign Office, Bank of England, universities, and local government. She is also something of a populist. Her views on capital punishment, immigration, and the trade unions resemble those of the right-wing tabloid press. Yet, to a greater extent than other mobilizers, say, Joseph Chamberlain or Lloyd George, she has operated within the established political institutions of Parliament and party.

In Britain, the character of being a mobilizer or a reconciler is not related to being in any one party or being in opposition or government. The movement between the two has often been cyclical, with one style breeding a reaction in favour of the other. A leader may start out as a mobilizer but end as a conciliator. In 1964 Mr Wilson presented himself as a radical figure, but by 1974 he was comparing himself to a family doctor, and hoped, in his own words, 'to achieve . . . peace and quiet for the country'. Mr Heath in 1970 was also a radical leader and many of his policies were enacted in the face of bitter opposition in and out of Parliament. Yet by October 1974 the once radical Mr Heath was espousing the cause of conciliation and national unity, arguing that the grave problems facing the country and the necessary mobilization of consent required a cross-party coalition.

Mrs Thatcher, however, has remained a mobilizer; she clearly regards herself as Heath plus, providing greater political will and persistence. Talk of a policy U-turn or reversal, associated with Mr Heath from 1972 onwards, or of 'ungovernability', so prevalent in the 1970s, has disappeared from the political vocabulary in Britain. Mrs Thatcher prides herself on her reputation as an 'Iron Lady'. But she is also reacting against a political 'fixer' style of leadership associated with the

Labour Prime Ministers Mr Wilson and Mr Callaghan in the mid-1970s. Philip Williams has drawn a distinction between Labour party leaders who are *path-finders*, who look for new solutions or try to lead the party in new directions, and *stabilizers*, who are concerned above all with preserving party unity.[22] Labour, with its well-entrenched factions, egalitarian values, myths about the sovereignty of the party Conference, and separation of powers between the National Executive, Conference, and parliamentary party, has required coalition building skills of a high order. Both Wilson and Callaghan were determined to keep the fissiparous Labour party together. They thought that this required the abandonment of any directive or heroic leadership and the careful balancing of rival right and left political factions in the Cabinet. Critics inveighed against the fudging, the soothing formula, the trimming, and the unheroic style, all designed to keep the party together. The success of the mobilizer is to be measured in the accomplishment of goals; conciliators are more interested in unity and consensus—even if it is directionless.[23]

It is always difficult to assess the political significance of an individual leader. One approach is to imagine different outcomes in the past and ask 'what if?' For example, what if Mr Heath had won the February 1974 general election, as he nearly did, and carried on as Prime Minister? What if Mr Whitelaw, the heir–apparent, had stood on the first leadership ballot in 1975 and won, as many think he would have? Even to pose such questions reminds us that there was a large element of chance in the emergence of Mrs Thatcher. Circumstances, or 'luck', have also been important in her later career, particularly the outbursts of industrial unrest in 1979 which so damaged the Labour government, the divisions in the Labour party, the occurrence of the Falklands war in 1982, and the divided non-Conservative vote in the 1983 and 1987 elections.

It is reasonable to associate a number of distinctive policies with the presence of Mrs Thatcher—the economic strategy, attempts to contain the public sector, toleration of high unemployment, reduction in the powers of trade unions, privatization, the vigorous prosecution of the war in the Falklands, the community charge, education reforms, change in the civil service, and Britain's refusal to join the EMS. A number of

Cabinet colleagues had doubts about all these policies and many would acknowledge that things would have been different without her.

It is doubtful that she has transformed the role of the Prime Minister in British politics. The office is not highly institutionalized and she has not made changes in this respect. Her significance as Prime Minister is that she has set an example; by pushing to the outer limits of her authority, making decisions with small groups of ministers and advisers, and closely involving herself in Whitehall promotions and the policies of departments, 'the repertoire of Prime Ministerial tasks has been extended'.[24] The very forcefulness with which Mrs Thatcher has projected her views and style, separate from those of her Cabinet, has also established a model of premiership. It is one which, whether measured in terms of winning elections or carrying through policies, has been remarkably successful.

Thatcherism has been both a matter of style—combining impatience with much of the status quo and a relentless promotion of new attitudes—and policies. Both Lloyd George and Churchill were helped by the fact that their style suited the temper of the times. Historians may point to the downfall of Lloyd George in 1922 and Churchill in 1945; the achievements were recognized but declined in importance as the electorate responded to new issues. Mrs Thatcher's achievements in taming the trade unions, reducing inflation, introducing privatization and more commercial values into the public sector, and recovering the Falklands gave scope for her adversary talents. As past achievements are taken for granted and new challenges emerge, so voters may look for alternative leaders.

Mrs Thatcher's style of leadership has thrived on crises. In such circumstances, the identification of an unpopular target —an Argentine General, Irish hunger strikers, or coal miners' leader Arthur Scargill—polarizes public opinion. By being clearly against an unpopular figure, Mrs Thatcher has usually rallied public opinion to her side. The same style is less suited to 'normal' politics or when the 'enemy' is less obviously unpopular. The support she may win in the crisis, where minor disagreements are sunk in the face of a much disliked enemy, can evaporate in everyday disputes, where the Prime Minister's views must compete with many other acceptable views.

Historians are certain to acknowledge that she dominated the political landscape of the 1980s and a number of her policies will leave a mark in post-Thatcher Britain. The success is even more remarkable when compared to her predecessors. With the possible exception of Baldwin in 1937 no Prime Minister in the twentieth century has left office willingly and on a high note. They have been overthrown (Asquith, Lloyd George), stepped down because of old age, ill health, or policy failures (Balfour, Bonar Law, MacDonald, Chamberlain, Churchill, Eden, Macmillan), or were repudiated by the voters (Attlee, Heath, Callaghan). It remains to be seen how, and indeed whether, Mrs Thatcher is able to manage her departure from office.

Notes

1. I have written about this in 'From Gentlemen to Players', in W. Gwynn and R. Rose (eds.), *Britain: Progress and Decline* (London, Collins, 1979), p. 32.
2. R. Crossman, Introduction to W. Bagehot, *The English Constitution* (London, Watts, 1964), p. 57.
3. A. Brown, 'Prime Ministerial Power', Parts I and II, *Public Law* (1968); and G. Jones, 'The Prime Minister's Power', *Parliamentary Affairs* (1965).
4. See A. Aughey, 'Mrs Thatcher's Philosophy', *Parliamentary Affairs* (1983) and Chapter 6 above.
5. Quoted in H. Young and A. Sloman, *The Thatcher Phenomenon* (London, BBC Publications, 1986), p. 14.
6. *The Thatcher Phenomenon*, p. 42.
7. See *The Times*, 19 January 1984.
8. *The Observer*, Interview, 25 February 1979.
9. See H. Young, *One of Us* (London, Macmillan, 1989).
10. A. King, 'Margaret Thatcher: The Style of a Prime Minister', in A. King (ed.), *The British Prime Minister* (London, Macmillan, 2nd edn., 1985), p. 98.
11. D. Hurd, *An End to Promises* (London, Collins, 1979), p. 32.
12. Sir Alan Walters (as he now is) provides a perspective on this experience in his *Britain's Economic Renaissance: Margaret Thatcher's Economic Reforms 1979–1984* (Oxford University Press, 1986).
13. See G. Jones, 'Cabinet Government and Mrs Thatcher', *Contemporary Record*, Autumn 1987.
14. On this see P. Hennessy, *Cabinet* (Oxford, Blackwell, 1986). For a different perspective see D. Willetts, 'The Role of the Prime Minister's Policy Unit', *Public Administration* (1987).

15. See ibid., pp. 96, 98.
16. R. Rose, 'British Government: The Job at the Top', in R. Rose and E. Suleiman (eds.), *Presidents and Prime Ministers* (Washington DC, AEI, 1980) and G. Jones; 'Presidentialism in a Parliamentary System', unpublished paper London School of Economics 1988.
17. A. Gamble, 'Thatcher—make or break', *Marxism Today* (November 1980), p. 17.
18. See P. Byrd, *British Foreign Policy under Thatcher* (Oxford, Philip Allan, 1988).
19. In 1979 the two main party leaders gained over 60 per cent of the time devoted to their political parties on BBC and ITV coverage of the election. No other politician reached double figures. In 1988 Mrs Thatcher gained over 60 per cent of her party's coverage and Mr Foot over 50 per cent of his party's coverage. In 1987 Mrs Thatcher gained 46.5 per cent of the party's coverage and Mr Kinnock 56.7 per cent of Labour's coverage. These figures are drawn from David Butler and Dennis Kavanagh, *The British General Election of 1979* (London, Macmillan, 1980), p. 209, *The British General Election of 1983* (London, Macmillan, 1984), p. 160, and *The British General Election of 1987* (London, Macmillan, 1988), p. 144.
20. R. Rose, 'British Government', pp. 20–1.
21. M. Cockerill, P. Hennessy, and A. Walker, *Sources Close to the Prime Minister* (London, Macmillan, 1984), p. 191. Also see M. Cockerill, *Live From Number 10* (London, Faber, 1988).
22. P. Williams, 'Changing Styles of Labour Leadership', in D. Kavanagh (ed.), *The Politics of the Labour Party* (London, Allen & Unwin, 1982).
23. R. Rose, 'The Variability of Party Government', *Political Studies* (1969).
24. A. King, 'Margaret Thatcher', p. 137.

10

THE REORDERING OF BRITISH POLITICS

THIS last chapter evaluates, under six headings, the significance of the Thatcher governments in post-war British politics. First, it discusses the government's impact on the Conservative party: has it created a new Conservatism? Second, it analyses the changed relationships between the government and major interests and pressure groups. Third, it analyses the influence on the political agenda and public opinion; has it shifted either of them to the right? Fourth, it compares the British experience with trends in other western states in the 1980s; has the so-called Thatcher revolution been no more than the British reflection of trends apparent elsewhere? Fifth, it examines whether recent elections have produced a realignment in the party system; have three successive handsome election victories now made the Conservative party the natural majority party? Finally, it assesses the impact of party government in Britain in the light of the Thatcher record.

A New Conservatism?

Politicians live by rhetoric. They often use words as tools for influencing public opinion, exaggerating or belittling policy achievements and even substituting for them. The Thatcher record since 1979 has been accompanied by a good deal of inflationary rhetoric from both supporters and critics. Admirers claim that the Thatcher governments have vanquished much conventional wisdom of the 1970s, such as the idea that the British were 'ungovernable', that the unions ran the country, or that there was a 'British disease'. Instead, it is claimed, Britain's economic decline has been reversed; in the process much of the post-war consensus has been broken, and a good thing too. On the other side, critics (including some Conservatives) claim that in their zeal to break the consensus the governments have destroyed much of manufacturing industry, starved the welfare state of resources, and divided British society. More detached observers suggest that in the next century historians may see Mrs Thatcher as a figure having no more long-term significance than Wilson, Heath, or Callaghan. One sceptical commentator has reacted against the idea of there being something called 'Thatcherism', by claiming that 'both opponents and supporters of the administration have created more of a pattern from disconnected events and policies than is warranted'.[1]

One approach to debunking the distinctiveness of the Thatcher record is to argue that the post-1979 policies are merely a continuation of earlier trends, for example, of the early economic policies of Mr Heath's 1970 government or of the acceptance of monetary targets and abandonment of Keynesianism under the Callaghan–Healey regime in 1976. There is, however, a distinction between the Thatcher policies and those of Mr Heath, which were short-lived, and those of Mr Callaghan and Mr Healey, which were reluctant, partly forced by outside pressures and largely confined to monetary targets and cuts in public spending. To claim that the government's record since 1979 has been the product only of instinct or opportunism, displaying no coherence, is, I would suggest, to carry scepticism too far.

It is true that policy blueprints are rarely carried through perfectly. Qualifications and amendments have to be made to the plans of the drawing-board as policy-makers confront changing circumstances. Yet doubts about political patterns may be overdone and result in a denial of such phenomena as a New Deal in the United States in the 1930s, Gladstonian Liberalism in the late nineteenth century, or collectivist politics after 1945. The setting of particularly high standards for uniformity and internal consistency makes it impossible to speak of the existence of almost any 'ism' in policy.

In terms of political style and policies there is something new and radical about the Conservative party under Mrs Thatcher. The trade union reforms, the break with the 'social partners' approach in running the economy, the abolition of a tier of local government, the imposition of far-reaching central government controls over local government finance and policies and then the replacement of household rates by the community charge, and the large-scale privatization, all mark a major departure from the conventions of post-war politics. The reactions of such leading Conservative figures as Francis Pym, Sir Ian Gilmour, Edward Heath, Jim Prior, and the late Lord Stockton, show that Thatcherism has offended many traditional Conservatives. Their criticism is not simply the embittered jealousy of defeated leaders; it represents a feeling of betrayal.

For most of its history the Conservative party has managed to avoid the factionalism of the Labour party.[2] The leadership of Mrs Thatcher and her repudiation of some of the policies associated with her Conservative predecessor certainly sharpened divisions among Conservative MPs in her first term of office. Those MPs of a so-called 'dry' tendency have regarded the control of inflation via a tight rein on the money supply as the primary objective of a government's economic policy, favoured free-market policies and a reduction of state ownership and spending, and taken an authoritarian line on law and order and many social issues. The so-called 'wets' favoured an interventionist role for government in the economy and were prepared to increase state spending to boost employment, even if this risked some increase in inflation. But over time the Conservative party divisions on economic policy became less marked, partly because of electoral success and partly because unemployment has fallen steadily from June 1986. It could fairly be said that Mrs Thatcher won the economic argument.

The simple division between 'dry' and 'wet' does not, however, do justice to the currents of opinion in the contemporary Conservative party. One of the effects of the Thatcher decade has been to redefine the benchmark of divisions.

The new Tory right consists both of libertarians and traditionalist (or Burkeian) Tories. These two groups agree on rejecting many economic policy prescriptions of the 'wets' but disagree on the role of the state. Libertarians want a drastic reduction in the interventions of the state, not only in the economy but also as a social regulator, so that opportunities for both personal choice and the free-market economy are maximized. Traditionalists, however, do not regard society as merely a gigantic market place and favour an authoritarian stand on many social issues, for example, on drugs, abortion, Sunday trading, and censorship. Norman Tebbit's Disraeli lecture in 1985 spelt out his distaste for the 'valueless values of the permissive society', of the 1960s and 1970s—represented by legalized abortion and homosexuality, fewer constraints on what is portrayed in the media and theatre, and growing disrespect for authority. By 1989 the largest Conservative back-bench group was the right-wing 92 Group.

A reading of Rhodes Boyson's *Centre Forward* (1978) gives some idea of how the Conservative party's policies have shifted in the past decade.[4] Boyson, an outspoken Conservative MP for Brent North, is on the right wing of the party and propounds populist authoritarian policies—on law and order, education, trade unions, and patriotism. In 1978 his views were shared by a small minority in the party. Boyson looked forward in his book to a seven-day programme for the reconstruction of Britain. The programme included:

(1) sharply reduced rates of income tax and a top rate of 60 per cent;
(2) government spending to be cut by 5 per cent annually in real terms for five years;
(3) a substantial increase in police numbers and pay;
(4) state welfare to be replaced by a scheme of payments and vouchers;
(5) an end to exchange controls and allowing the pound to find its own level;
(6) an end to the statutory monopoly of nationalized industries and all to be opened up to competition.

In 1978 these policies appeared far-fetched. Yet by 1988 substantial progress had been made on all but items (2) and (4) of the agenda.

Some of Mrs Thatcher's critics claim that she is not a Conservative at all, at least not in the paternalist tradition of Disraeli and One Nation Conservatism.[5] The qualification is important. Conservatism in Britain has been multifaceted but that tradition has been dominant. Her supporters, however, claim that she has returned the party to the essential verities of sound money, lower rates of income tax, strong government, and rolling back government from areas where it has no useful role to play. From this perspective it was Macmillan, Butler, and their heirs who betrayed Conservatism, by compromising with social democracy.

The case against Mrs Thatcher as a Conservative includes her impatience with the status quo and her suspicion, indeed rejection, of so many traditions and institutions. Much of the alleged wisdom that inheres in institutions and practice is part of what she has regarded as the flabby consensus that dragged the country down. She has not approached the senior civil service, local authorities, universities, trade unions, BBC, Church of England, the legal and medical professions, and public corporations with the reverence—or at least respect—with which Burke recommended that we regard the work of our ancestors.

Sir Ian Gilmour, a member of Mrs Thatcher's Cabinet until his dismissal in 1981, has been the most articulate critic of the new Conservatism. His views have been expressed in numerous speeches and newspaper articles and in two learned and witty books, *Inside Right* and *Britain Can Work*. He claims that Conservatism is 'not an idea, still less is it a system of ideas' and that the Conservative tradition is marked by 'scepticism, a sense of the limitations of human reason, a rejection of abstraction or abstract doctrines, a distrust of systems and a belief in the importance of experience and circumstances'.[6] He invokes Burke in his opposition to a priori politics and to the application of systematic and abstract schemes and policies. Conservatives should not be rationalists. The kind of ideological politics which he castigates as non-Conservative covers monetarism as well as Marxism. Francis Pym, who was abruptly dismissed as Foreign Secretary in 1983, is another good exemplar of the old consensual Tory style. He favours an approach which is 'balanced', aware of different points of view, and he is critical of Mrs Thatcher's '"absolutist" style—absolutely in favour of one thing, absolutely against another'. Lord Whitelaw, for long her Deputy Prime Minster, also embodied this kind of Conservatism—cautious, moderate, a classic One Nation Tory. As the product of a privileged background, he surely represented the type of rich Conservative who was inhibited by what Mrs Thatcher once called 'bourgeois guilt' (see p. 246), and a directionless consensus. In his memoirs he answers the charge: 'Some Tories, like me, are criticised for this supposed failing (of being patrician) which is said to lead to moral softness, in contrast with the toughness of those who have had to fight their way in life. I am not ashamed to be regarded as sympathetic to the anxieties of those whom life appears to be treating harshly. Nor do I accept that such feelings are a sign of weakness, provided that they are accompanied by an inner toughness where necessary and justified.'[7] He told Brian Walden, in *The Sunday Times* (16 April 1989), 'I don't pretend that I liked her because I didn't . . . But I decided to force myself to try to get on with her. I found her infuriating . . . she goes on and on and is determined to get her own way.' All three are the voices of the Tory grandees mourning the demise of their values under the shopkeeper's daughter.

In one significant sense, however, Mrs Thatcher has operated in the tradition of Conservative statesmanship. Jim Bulpitt argues that an abiding interest of Conservative leaders has been the protection of the autonomy of the central government in matters which ministers regard as 'high politics', such as defence, foreign affairs, and national economic policy.[8] Other issues—of 'low politics'—could be left to voluntary groups, the market, or local government. By the 1970s, critics claimed that government had become 'overloaded' with responsibilities and popular expectations—such as maintaining full employment or delivering rising living standards. In economic policy, of course, this autonomy of government had been further compromised by Mr Heath's search for agreement with the trade unions on incomes policy between 1972 and 1974 and by succeeding Labour governments.

The task therefore was to recover greater autonomy at the centre and limit the state's obligations to those that were essential and/or that it could carry out. As already noted, the attraction of monetarism for the new Conservative leadership by 1979 was that it promised a method of controlling inflation by means, as it were, of an automatic pilot; there would be no need for government to compromise its authority with producer groups. Strict control of the money supply and a curb on government borrowing would squeeze inflation out of the economy. The government would not print the money to pay for inflationary wage deals. Instead it would land trade unions and employers with the consequences of their decisions on prices and pay. Abandoning the commitment to full employment would restore autonomy to the centre. According to Bulpitt, 'in so far as monetarism was associated . . . with "arms' length", anti-corporatist government, then . . . the statecraft (of monetarism) must be viewed as less a radical break with the past and more as an attempt to reconstruct it'.[9]

This view can be related to a number of outstanding features of the British political system. Regardless of whether it is called a 'top down' model or an elective dictatorship,[10] the formal concentration of political authority in Britain is remarkable. A unitary system, the sovereignty of Parliament usually wielded by a one-party majority, Treasury control of national finance, the absence of a written constitution or a Bill of Rights which courts can use to challenge the government, and the idea that extra-parliamentary party organs are merely 'handmaidens' of the parliamentary party all bear witness to this centralization. It has proved convenient for Conservatives and Labour parties and will probably continue to be so as long as one of them holds office. The central issue in British politics has not been how to curb the elective dictatorship but how to capture it.

For all the Conservative talk of 'rolling back the state' there has been a remarkable centralization of decision-making since 1979. To understand this, it is necessary to distinguish between the areas of defence, law and order, and social policy—where the motive of government policy appears to be the concentration of authority to build a 'strong state'—and full employment, incomes policy, industrial policy, and public ownership from which the government has withdrawn to create 'a free economy'.[11] What is significant is the extension since 1979 of the range of issues in the first category which the Conservative leadership regards as 'high politics', that is, which it regards as being within its own sphere of decision-making. In education, local government, and health the government has intervened and regulated on a large scale.

The Interests Disturbed

Traditionally, Conservatives have been respectful of the claims of voluntary groups, interests, and self-governing professions; these were Edmund Burke's celebrated 'little platoons'. By imposing new policies and challenging traditional methods of decision-making, however, the government has disturbed a number of interests. In higher education, for example, the idea of the autonomy of the University Grants Committee (replaced in 1988 by the University Funding Council) as a buffer between the universities and the Department of Education and Science has been shattered and the Department has regularly intervened with advice to universities and polytechnics about subject areas which should have greater or less resources. In secondary and primary education the government has imposed central pay bargaining on the school teacher unions, a contract of service and appraisals of performance on teachers, and a core curriculum. The measures represent a major coup by central government against local government and the unions. It massively increased the budget of the Manpower Services Commission, enabling it to play a dominant role in vocational education.

In the National Health Service the local element in policy-making has been weakened as the government has pushed for greater accountability of regional bodies to central government. The Secretary of State for Health reviews proposals to promote efficiency with each Regional Health Authority chairman and the latter then meets the chairmen of the District Health Authorities to implement the policies, covering specific spending and manpower targets. In 1989 the government's proposals for the reorganization of the NHS included the creation of an NHS 'internal market' in which GPs could shop around for health care, the establishment of a new management board, with most members drawn from industry, permission for hospitals to become self-governing and new contracts for GPs. The 1983 Water Act reduced local authority representation on Regional Water Authorities and the new boards have fewer members, all of whom are now appointed by the Environment Minister. As with education these interventions and the reduction in the role of producers and local government spokesmen have usually been made in the name of promoting value for money and increasing choice for consumers. The search for efficiency and cost-cutting has resulted in greater centralization.

The most remarkable extension of central control has been in the field of local government. For long British Conservatives praised local government and voluntary groups as a check on an over-mighty (obviously Socialist) central government; such pluralism could also promote experimentation and diversity in policies. The party, for example, has always opposed mandatory comprehensive secondary education, insisting that the choice be left to each local authority. Labour, which wanted mandatory comprehensive schooling, was painted as the party of central control and uniformity.

Conservative defence of local government prevailed until the mid-1970s. Thereafter, the growing concern over inflation, steep rises in household rates, growth of public spending, and the activities of left-wing Labour authorities led the Conservatives to search for economy. This culminated in the imposition of far-reaching controls. The containment of public spending has won out over local choice; reducing the state in this area has meant central government rolling back the local state.

The erosion of consensus politics overtook local government as it did many other areas of public life. In Manchester, Liverpool, the Greater London Council, and some London boroughs left-wing Labour authorities sought to implement local socialism and use the town hall to challenge Thatcherite Whitehall. Local defiance of central government policies and rejection of its ultimate authority overturned traditional conventions of central-local relations. Increases in spending, grants, and support for anti-racist, pro-Sinn Fein, and pro-gay-lesbian policies enraged right-wing Conservatives and the tabloid press. The label 'loony left' councils stuck and was used to discredit much of local government in general. There also developed a right-wing Conservative critique of local government. So often, it was claimed, local government was not the voice of the local community—election turnouts were derisory, elections were fought on national rather than local issues, and many voters were council employees (with, therefore, a vested interest in greater spending), council house tenants (with an interest in subsidized rents), or were not ratepayers at all. Many of the services provided were quasi-monopolistic and poor.[12] The high spending, high rating left-wing councils drove away business, exacerbated the economic decline of inner cities, and frustrated the enterprise policies of the government.

Increasingly the Thatcher governments became the protector of the ratepayer rather than the defender of local democracy. According to one commentator Thatcherite decentralization embraced a new 'ratepayer democracy':

> Local government itself was seen as being as much in need of a rolling back of its frontiers as was central government, in order that economic individualism might have greater room for action. Ratepayer democracy . . . implied less local government, with fewer services to provide since some of them could be handed over to the private sector, to the supposed benefit of consumers and ratepayers alike.[13]

The attempt to 'discipline' certain offending local authorities involved the government in many political and legal battles, the passage of several Acts of Parliament, and, ultimately, the creation of a radically reshaped system of local government. In the 1980 Local Government Planning and Land Act the government took powers to decide the spending totals of each local authority. When this failed to contain local spending sufficiently, volume targets were introduced together with penalties in the form of loss of central government grant if local authorities exceeded their targets. In response defiant authorities, invariably Labour controlled, raised the extra money from household and business rates. Having tried and failed to control spending the government then decided to control local revenue. The 1982 Local Government Finance Act abolished the right of authorities to set a supplementary rate and the 1984 Rates Act empowered the government to limit the spending and rates of centrally defined overspending authorities. Finally, in some frustration, in 1988 the rating system was abolished for England and Wales and a community charge introduced, to take effect in 1990–1 (introduced a year earlier in Scotland). Whereas only about a third of local residents paid full rates, most adults will pay the charge. By restoring the link between voting and taxation, the government hopes that voters will punish 'wasteful' or 'high-spending' councils. In 1985 legislation was also passed to abolish the GLC and the metropolitan counties (all of which were Labour controlled). By 1987 Britain was the only European state without a directly elected capital city authority.

Some commentators on the left take a more explicitly political view of the central-local government trends since 1979 and suggest that they are part of a deliberate larger attempted restructuring of relations between state and society. The abolition of rates fulfils an ambition of Mrs Thatcher's dating back to 1974. But the interventions, each one usually more radical than the last, were more a sign of growing frustration at the centre. Critics on the left claim that the police and education services have become part of the strategy of state control and regulation. This has been combined with such measures as the privatization of local housing and other services, central control over local spending and revenue, and forcing local authority direct labour organizations into the market place. Urban development corporations (UDCS), were established, dominated by private business, to promote the regeneration of inner cities and bypass local authorities. In education and housing the government has limited the role of local government. The centralization from above, and decentralization to markets and consumers below, has weakened local government.

The effects of large-scale unemployment and changes in the laws affecting trade unions have sapped militancy on the shop floor. The National Graphical Association incurred four contempt fines totalling £675,000 and eventual sequestration of its funds before it called off its illegal picketing (under the 1980 Act) against Mr Eddie Shah's publishing company in 1983. The National Union of Mineworkers also had to pay heavy fines and suffer sequestration. In 1986 Sogat 82 was fined and had its entire assets sequestrated because of its illegal disruption of the distribution of Rupert Murdoch's News International newspapers. Mrs Thatcher's governments were determined to cut the unions down in size and, for all the incremental nature of the legislation, prepared the ground in advance.[14]

Class analysis alone is inadequate for an understanding of the policies of Thatcherism, for it has also disturbed a number of middle-class interests. The ending of the monopoly of solicitors over house sales conveyancing and of opticians over the sale of spectacles has already been noted. A Conservative attack on such solid middle-class professions is unprecedented. The pace has increased in the government's third term of office. Tenure for university staff has been abolished. In 1989 it was planned that doctors would have to work to new contracts, under which their pay would be more clearly related to the number of patients. There are also government proposals for the reorganization of the legal profession. Barristers will no longer have the exclusive right of audience before higher courts; any lawyer possessing an 'advocacy certificate' will be able to appear. In 1989 the proposals were being bitterly resisted by doctors and lawyers.

Equally notable has been the worsening relationship with leaders of the Church of England—for so long known as the Tory party at prayer. Some bishops, notably Jenkins of Durham, Sheppard of Liverpool, and Hapgood of York, have spoken out about deprivation in the inner cities, the miners' strike, and the need for government to show a greater compassion for, and understanding of, the poor. According to the Bishop of Durham, 'This Government also seems to be indifferent to poverty and powerlessness. The financial measures consistently improved the lot of the already better off while worsening that of the badly off . . . such a Government cannot promote community.' According to the Archbishop of Canterbury, 'There seems to be a movement from consensus to confrontation and also a growing scale of confrontation.' The archbishop also called, in a newspaper interview in November 1984, for 'leadership in our national life which will unite and not divide the nation'. He added: 'That's not an attack on the government.'[15] Later, in December 1985, the Archbishop of Canterbury's committee on the inner cities presented its report, *Life in the City*. This contained many proposals for the regeneration of these areas, including greater central government expenditure, which ran counter to the government's ideas. The report was immediately dismissed as 'Marxist' by a government spokesman and criticized with varying degrees of vigour by many Conservatives.[16]

These attacks in turn have been met by forceful rebuttals from Conservative back-benchers and from the Prime Minister herself. The Tory critics object that the Church is peddling left-wing politics as a religious message, while failing to assert moral values. Mrs Thatcher in 1988 addressed the General Assembly of the Church of Scotland and quoted St Paul: 'If a man shall not work—he shall not eat.'

Relations between the government and senior civil service also have been uneasy. Some might regard the frequent leaks of sensitive government documents as a consequence of this tension. Mrs Thatcher also appears to have intervened more directly than previous Prime Ministers in the promotion of Permanent Secretaries, preferring less senior and 'obvious' candidates on the grounds that they were impatient with the status quo and not 'defeatist'. Some have accused Mrs Thatcher of applying tests of political loyalty in making appointments. In fact she has favoured 'doers'. Mrs Thatcher has been fortunate in that retirements of several senior personnel in the early 1980s gave her the opportunity to influence promotions; there were thirteen appointments as Permanent Secretary in 1982 alone and she has had a hand in appointing the great majority of Permanent and Assistant Permanent Secretaries since 1979.

The traditional 'trade-off' for the secure tenure of senior British civil servants has been that they are anonymous and politically impartial. To some degree the impartiality has depended on there being a large measure of policy agreement and continuity between the parties in government. The doubts of some pre-war Labour leaders that the civil service would actually co-operate with a radical Labour government were largely laid to rest by the experience of the Attlee government. But radicals in both parties who favoured radical changes, either to the left or to the right, have frequently condemned the civil service as the embodiment of an above-party consensus favouring continuity of policy.[17]

Sir John Hoskyns, a former head of Mrs Thatcher's Policy Unit, has called for a radical reform of the government system if the effort to transform the economy is to succeed. He urged the appointment of more outsiders to top positions in the civil service and of officials and ministers who were committed to radical change.[18] A rejection of this approach was voiced by Lord Bancroft, head of the civil service until 1981. In a lecture in 1983 he argued that a civil servant who could answer 'yes' to the question 'Is he one of us?' should retire and become a party politician.[19] Whether it was the abandonment in 1981 of established procedures for settling the pay and conditions of civil servants, abolition of the Civil Service Department in the same year, deterioration of levels of pay relative to the private sector, efficiency scrutinies in departments in search of savings, attempts to limit index-linked pensions, or the reduction in numbers of staff, there was no doubting Mrs Thatcher's hostility to much of the ethos of the civil service. In 1989 the government was preparing to implement *The Next Steps* under which many departments carrying out executive functions would be converted into independent agencies. Her government, by its clear breach with many previous policies and early derision of the higher civil service, has placed a double strain on the traditional relationship. She has been willing to look outside the civil service for advice on policy—to various 'think tanks' and individuals who have often reinforced her scepticism about traditional departmental policy views. The new emphasis is on management skills and giving value for money.

The decision in January 1985 of the Oxford University Congregation to vote by a margin of more than 2 to 1 against awarding its most famous daughter an honorary degree is another mark of the emotions she raises among other established élites. Opponents resented the cuts in university funding and complained that the government was hostile to higher education, particularly to studies of the humanities, which fostered a critical spirit of enquiry. The 'utilitarian' policy of shifting the balance in higher education to vocational, business, and technological subjects concerned many university leaders; it was memorably dismissed by Enoch Powell as 'barbarism'. Mrs Thatcher, and the attitudes that she presents, were a challenge to many established ways in the universities. Hugo Young ('Why the Dons Showed They've Learnt the New Politics', *Guardian*, 4 February 1985) appreciated this side of Mrs Thatcher and its relevance to the question of the honorary degree. He commented:

> Far from being a blot on her tenth anniversary as Conservative leader, the withholding of the doctorate perfectly crystallised the cultural change that she has wrought . . .

She has been as ready as the Labour party to apply political criteria when making appointments to bodies like the health authorities and nationalised industries.

She praises the risk takers, wealth creators, people who are frequently from modest backgrounds and who may not even have been at a university. Witness the speech at Newcastle on 23 March 1985, when she attacked her clerical and academic critics:

> And nowhere is this attitude [opposition to wealth creation] more marked than in the cloister and common room. What these critics apparently can't stomach is that wealth creators have a tendency to acquire wealth in the process of creating it for others.

She continued:

> They didn't speak with Oxford accents. They hadn't got what people call the 'right connections', they had just one thing in common. They were men of action.

Her desire to break with traditional policies has also involved a rupture with many traditional Tory interests and leaders. Before their deaths both Lord Butler and Lord Stockton, formerly Harold Macmillan, frequently made clear their distaste for the government's policies. These eminent figures from the age of consensus felt that Mrs Thatcher was betraying their 'One Nation' legacy. Although leader of the Conservative party she is pre-eminently an anti-establishment figure. It is interesting that her famous question 'Is he one of us?' has usually been asked about fellow-Conservatives.

One mark of a radical government is its politicization of hitherto relatively non-political areas. The list is long: the government's decision in August 1985 to lean on the BBC Board of Governors not to show a *Real Lives* programme featuring interviews with terrorists in Northern Ireland; its prohibition of broadcasts with terrorist groups in Northern Ireland; its prosecution of newspapers which revealed the contents of the *Spycatcher* book of memoirs by a retired MI5 agent, Peter Wright; its robust dismissal of the dissenting views of senior figures in the Church of England, senior civil servants, and university spokesmen; its appointments to various quangos and nationalized industries; and its controls on local government. All may have emulated some actions of previous British governments. But the broad package of interventions and centralization over such a long period of time surely outstrips that carried out by any previous government. In many ways it is a classic case of adversary politics. In so far as the actions since 1979 have politicized areas hitherto relatively immune to such considerations they provide pretexts for a future Labour government to intervene in these spheres.

Another casualty of Thatcherism has been what Whitehall calls 'The Great and the Good', those who often staffed the Royal Commissions and committees of inquiry. The G and G were an embodiment of the old consensus. They were largely drawn from academe, retired mandarins, and BBC governors; they were often political centrists and cross-benchers who had a belief in disinterested enquiry and collecting and weighing evidence. They were schooled in 'squaring interests' and coming up with acceptable compromises. Harold Wilson appointed so many commissions that he brought the system into discredit. There is some merit in Mrs Thatcher's claim that too often Royal Commissions were appointed largely to avoid decisions or to postpone necessary tough action: she has not appointed one Royal Commission in her ten years in office.[20] After all, she knows what is to be done. Her preferred method has been to make policies with small working parties, drawn from sympathetic ministers and members of her Policy Unit and 'think tanks', as with the introduction of the poll tax, the health service review, and some of the education reforms. Peter Hennessy has called the old G and G 'the lost tribe of British public life'.

How does one reconcile this record with the Burkeian respect for structures and institutions which mediate relations between the citizen and the state? In the 1980s Conservatives have been more likely to evaluate the mediating institutions according to whether or not they are in the public sector. If they are—including public-sector trade unions, local government, local authority schools and housing, and various quangoes—then they attract little sympathy and people have to be liberated from them. Hence the encouragement for voluntary schools, City Technology Colleges, and for schools and hospitals to 'opt out'; inducements for universities to be less dependent on state finance; promotion of council house sales and housing action trusts. It is also interesting to note how relations have worsened with a number of élite institutions which have been identified with paternalist and consensual values—universities, the Church of England, the senior civil service, the BBC and One Nation Conservatism. These were committed to a balanced or cross-bench political outlook and have been uneasy with the Prime Minister's zeal, certainty, and forcefulness.

It is now commonplace to say that the Thatcher governments have had to be highly interventionist in order to extend the market, increase the rights of consumers, and reduce producer power. It has reduced the claims that groups are able to make on the state. Reliance on monetarism rather than incomes policy meant that the government no longer had to compromise its authority in making social contracts and wages policies with producer groups, particularly the unions. The public sector of the economy has been rolled back and strikes in the public sector have been resisted. Another tactic has been to give greater scope to consumers *vis-à-vis* public sector groups. Schools and housing associations can opt out of local authority control. A greater voice has been given to parents, governors, and head teachers in state schools and to GPs and patients in the health service. Trade union members have been given more say via pre-strike ballots, regular elections of their officials, and votes on the unions' political levy. Many nationalized industries, which long posed problems for governments, not least in pay negotiations, were floated to the private sector. The losers have been local authorities, trade unions, health authorities, and nationalized industries. The net effect of these measures has been to give greater autonomy to the central government.

Public Opinion

What effect has the government had on the public mood? One has to recall that a major objective of Mrs Thatcher's government has been to change public opinion, not least about the role of government itself. Mrs Thatcher has said that if economics is the means, then changing hearts and minds is the goal. One might think that such an ambition—no less than the forging of a cultural revolution—is most unconservative. Yet this is a natural task for Margaret Thatcher; she relishes proclaiming Thatcherite values and telling people which policies are morally right and which are morally wrong. Indeed the long-term success of free-market policies depends on a change in public mood. In Zurich in 1977, Mrs Thatcher said:

In our philosophy the purpose of the life of the individual is not to be the servant of the state and its objectives but to make the best of his talents and qualities. The sense of being self reliant, of playing a role within the family, or owning one's own property, of paying one's way, are all part of the spiritual ballast which maintains responsible citizenship . . .

The speeches of ministers have reiterated the themes that it is not the job of government to solve as many problems as previously, that ministers should be more attentive to the interests of taxpayers when spending public money, that 'real' jobs will be created and sustained not by government subsidy but by workers making goods which people will buy, that the criterion of 'value for money' be applied to public-sector activities, and that the private sector should be encouraged because it creates the wealth which the public sector requires. Conservative spokesmen have said that people should reduce their expectations of what government can do and that economic improvements would depend largely on the action of people themselves. In a word, the dependency culture should be replaced by an enterprise one. In shifting responsibility from government to the people: 'The Thatcher government has adopted a novel approach. It has tried to change the question'.[21] By the end of the year, in an interview with the *Financial Times*, Mrs Thatcher suggested that a third successive Tory election win would fulfil her aim of 'killing off Socialism'.

Surveys suggest that the government has had some success. A MORI poll (April 1989) for a BBC programme, *Thatcher's Children*, found that among young voters (15–28) aspirations were distinctly bourgeois. Their ambitions were to own their own homes and have private pensions. A remarkable 40 per cent wanted to run their own business and a quarter expected to do so. In 1979 and in 1983 there was clear electoral support for several Conservative policies, even among many Labour voters. On such issues as public ownership, legal curbs on the trade unions, defence, council house sales, and restricting the range of welfare benefits, a majority of Labour voters disagreed with their party's policies. By 1989 Labour had shifted its policies on many of these issues. In this sense we may talk of there being a shift to the right, even where this is not reflected in voting for the Conservative party.

There has also been a change in mood about nationalization. The 1964 British Election Survey found that just over half (51 per cent) were satisfied with the status quo, and supporters of further nationalization outnumbered denationalizers by 28 to 21 per cent. By 1979, however, the survey found that there had been a major shift in support for privatization and public opinion remained in favour of further privatization in 1983, and probably reflected a positive response to the Thatcher policies. By 1987, however, most voters, supported the (changed) status quo and were opposed to plans to privatize electricity and water.

Surveys have reported heavy support for the trade union reforms introduced by the Conservative government since 1979 and show that profit-making is still socially acceptable and that people endorse the principle that wage increases must be earned by firms selling their goods at a profit. In terms of ideology also, the Conservatives are well placed. In 1983, for example, Gallup found that of voters who placed themselves on the political left and right, 22 per cent placed themselves at various ranges of the left end of the scale (from 'far' to 'slightly' left), 51 per cent placed themselves on the right, and 13 per cent regarded themselves as middle of the road. In 1987, the respective figures had changed to 28 per cent, 45 per cent, and 9 per cent. Although there had been a 6 per cent swing from 'right' to 'left' between the two elections the former were still in a clear majority. Many Tory party cheer-leaders boast that there has been a cultural revolution. Conservative electoral landslides, extension of home ownership, reductions in trade union membership and power, and the rise of white-collar employment, entrepreneurship, and self-employment all seem to be powerful indicators of this change.

On the other side sceptics point out that the Conservative share of the vote in the 1983 and 1987 elections is below that in the 1960s and they point to the several opinion surveys which show that popular support for collective values remains high. In spite of undoubted social and economic changes there has been no revolution in values. As a careful analysis of the data

TABLE 10.1 *Attitudes towards Taxes and Benefits 1978–1989*

Question: People have different views about whether it is important to reduce taxes or keep up government spending. How about you? Which of these states comes closest to your own view? (Figures as percentages of those asked).

	Oct. 1978	May 1979	Mar. 1983	Feb. 1985	May 1987	Apr. 1989
(*a*) Taxes being cut, even if it means some reduction in government services, such as health, education, and welfare	25	34	23	16	12	10
(*b*) Things should be left as they are	23	25	22	18	21	14
(*c*) Government services such as health, education, and welfare should be extended, even if it means some increase in taxes	39	34	49	59	61	73
Don't Know	13	7	6	6	6	3
Net support for spending over tax cuts	14	0	26	43	49	63

Source: Gallup.

states, 'Quite simply there has been no Thatcherite transformation of attitudes or behaviour among the British public. If anything the British have edged further away from Thatcher's position as the decade has progressed.'[22]

Notwithstanding the re-election of Mrs Thatcher in 1983 and 1987, a clear majority of voters have favoured increased taxes to pay for greater expenditure on services such as health, education, and welfare, over tax cutting and programme cutting (Table 10.1). Since 1979 the balance has moved steadily against tax cutting in favour of more spending on education and the NHS. In May 1979 spenders and tax-cutters were equal (34 per cent); by May 1987 the spenders outnumbered the cutters by more than 5 to 1 (61 to 12 per cent). The results have not been dissimilar when Gallup has asked whether voters preferred a cut in local rates if this meant a reduction in local services or an extension of the services even if this involved an increase in rates.

In 1977, before Mrs Thatcher came to power, Gallup found that 33 per cent of voters thought that if people were poor this was probably due to their own lack of effort. By March 1985 the figure had fallen to 21 per cent. The proportion blaming poverty on 'circumstances' rather than fecklessness rose from 30 to 49 per cent between the two dates. The same survey found that three-quarters of the electorate felt that money and welfare should be more evenly distributed in Britain. Over time the proportion preferring 'caring' to 'tough' policies has increased by one-third and the 'carers' are now in a substantial majority.

The 1983 British Election Survey also found a change over recent years in favour of 'liberal' social policies. Although support for some of the liberal values is confined to a minority, it is a growing one. For example, between 1966 and 1983 support for abolition of the death penalty increased from 18 to 37 per cent (support for its retention or restoration fell from 82 to 63 per cent). A similar shift occurred on attitudes to policies which promote equal opportunities for women and ethnic minorities and the provision of abortion on the National Health Service. In all three cases there has been a marked increase in the proportion who favour the more liberal status quo.[23]

Public opinion has shifted to the right over the past decade on one component of collectivism—nationalization—and in a more liberal direction on another, state provision of welfare and social policies (for similar trends elsewhere, see below). If anything, the surveys suggest that the public mood was at its most Thatcherite and right-wing in the second half of the 1970s.

This divorce between the two elements is particularly marked among the salariat—the well-educated members of the professions and management—which has grown most in the past twenty years. The shift to a more 'liberal' view on social issues and disarmament is probably due to the growth of this group and the expansion of higher education.[24]

A decade has not made the concept of Thatcherism attractive to the British people. According to Gallup (April 1989) the term Thatcherism has negative connotations for most people. MORI (April 1989) found widespread support for the idea of a

TABLE 10.2 *'Collective' versus 'Individual' Values* (per cent)

Ideals: People have different views about the ideal country. In each of the following alternatives, which comes closely to the ideal for you and your family?

	All	Conservatives	Non-Conservatives
A country in which private interests and a free market economy are more important	30	46	21
or			
A country in which public interest and a more managed economy are more important	62	48	69
A country which emphasizes the social and collective provision of welfare	54	34	65
or			
A country where individuals are encouraged to look after themselves	37	61	25
or			
A country which emphasizes increasing profitability even if this means people losing jobs	29	44	21
A country which allows people to make and keep as much money as they can	48	65	40
or			
A country which emphasizes similar incomes and rewards for everyone	43	29	50
A country which emphasizes increasing the standard of living more than protecting the environment	25	24	26
or			
A country which emphasizes protecting the environment more than increasing the standard of living	61	63	60

Source: MORI April 1989.

society which placed collective over individual values (Table 10.2). Surveys of attitudes to so-called Thatcherite ideas (for example, a government sticking to its principles, however unpopular; not consulting with the major economic interests; and the government's inability to do much about unemployment) found that, on average, Thatcherite supporters (40 per cent) were outnumbered by opponents (47 per cent). Table 10.3 shows a further swing away from Thatcherite ideas towards caring and consensus government between 1983 and 1987. It is as if the public recognizes that society has changed over the decade but does not much like what it sees.

TABLE 10.3 *Support for Thatcherite Beliefs about Politics* (per cent)

	1983		1987		Surplus of Pro over Anti	
	Pro	Anti	Pro	Anti	1983	1987
When dealing with political opponents, stick firmly to one's beliefs (agree), or meet halfway?*	50	39	45	48	11	−3
Government can't do much to create prosperity; it is up to people to help themselves (agree)	48	38	39	48	10	−9
In difficult economic times, should government be tough (agree) or caring?	46	35	36	50	11	−14
In dealing with the rest of the world, is it better for Britain to stick resolutely to its own position (agree), or meet other communities halfway?	29	59	28	63	−30	−35
When governments make economic policy, is it better to keep unions and business at arms length (agree) or involve them?	27	62	22	69	−35	−47
Average	40	47	36	55	−7	−19

*Excludes responses which are mixed or Don't Know.

Agreement with statement indicates support for Mrs Thatcher's beliefs

Sources: BBC Gallup Survey, June 1983; Gallup, June 1987.

The burden of the survey findings seems clear and is rather disheartening for the Thatcher project. But some qualifications are in order. What are many of these questions measuring and how significant is the data, say as a guide to behaviour or as an influence on voting? Huge majorities apparently prefer reducing unemployment to fighting inflation. Yet what do we then make of a Gallup finding (21–6 May 1987) that, when asked to choose between inflation and unemployment as the greater threat to them and their families, 49 per cent chose inflation and only 43 per cent unemployment? Similarly, for all the huge majorities claiming to prefer more government spending, even if this involves paying more taxes, what does one make of the finding that voters, when asked *which mattered more to themselves and their families*, 'more public spending or tax cuts?', a majority preferred the tax cuts (Marplan, March 1987). Gallup has found widespread dissatisfaction with the quality of society. People are seen as being more selfish, aggressive, and less tolerant and happy than was the case a decade ago. But, surprisingly, when individuals were asked whether they felt this about themselves only tiny minorities admitted to such a feeling. Survey questions pitched at a general level may elicit what respondents regard as socially approved values.

Similarly worded questions over time have to take account of changes in the topics under question. Surveys since 1979 show that trade unions have become less feared and that there is less support for further privatization. On both topics, however, the status quo has changed radically compared to 1979. Unions have lost powers and the numbers of disputes has declined; similarly the public sector has been reduced in size. The data might be read as a backhanded success for the Thatcher government. As the concerns of 1979—high taxation, trade union power, nationalization, and inflation—have become less salient, so there is less support for radical policies to deal with them. As economic satisfactions have increased, so the scope for dissatisfaction on social issues may also have increased.

A New Middle Ground

There have been many claims that Mrs Thatcher has shifted the agenda of British politics. On the political left a journal like

Marxism Today claims that there has been an underlying and probably permanent shift to anti-socialist values. Again, Andrew Gamble is worth quoting:

> Whatever the fate of the Thatcher government, the New Right will survive it. In its impact on British politics and economic policy it has already been remarkably successful. The traditional post-war argument between different kinds of interventionism has been replaced by a much broader debate.[25]

The 1983 and 1987 Gallup election surveys showed that voters across all parties agreed on many goals of government. This is significant, given that many commentators in 1983 thought that the policies of the Conservative and Labour parties were further apart than at any time over the previous thirty years. Richard Rose's review of the Gallup survey on twenty political issues found that majorities of voters for all three parties agree with each other on more than two-thirds of all issues, that most Alliance voters agree with the national majority on nineteen of twenty issues, and that most Conservative and Labour voters with the majority on seventeen of twenty issues.[26] Moreover this level of inter-party agreement has persisted since 1970. Even in 1987 voters did not expect the return of a Labour or Conservative government to make a significant difference. According to Gallup, voters expected that, on twelve issues, a change of government would produce different outcomes only on those affecting trade union powers and nationalization. There is also wide support for the mixed economy—a private enterprise economy, subject to government controls—as well as for state provision of welfare services. The roles of government and state are viewed neither in the negative terms of Mrs Thatcher nor in the positive terms of Labour's left wing.

For all the talk about a new consensus and a new realism, however, a number of issues remain open. How is the free-market economy to be reconciled with continued large-scale tax concessions for house mortgages and private pensions? How durable, for example, will be the changes in industry, elimination of many restrictive work practices, productivity gains, decline of labour militancy, and reassertion of managerial authority? In 1989 the British economy was again at the wrong end of league tables for having relatively high levels of inflation, interest rates, and a trade gap. Some critics suggested that many of the old problems of the British economy had returned.

Yet if talk of permanent or irreversible changes may be too bold, the Thatcher government has created a new agenda, one which a successor government will find difficult to reverse. It is difficult to imagine that there will be a return to the status quo ante regarding the closed shop, trade union immunities and pre-strike ballots, the right of council house tenants to buy their own homes (over a million former council tenants have made such purchases), and the position of former nationalized industries. Millions of voters have acquired a stake in the wider ownership of shares and homes and a voice in union affairs. Similarly, sustained government rhetoric about getting value for money from programmes and protecting the interest of taxpayers will probably leave a mark. Since 1979 there has been a loss of over 1.5 million manufacturing jobs; by 1987 manufacturing accounted for only 22 per cent of the British workforce, 'a figure more typical of countries with large fishing or agricultural sectors'.[27] That decline, when coupled with the abolition of Labour party strongholds in local government in London and the metropolitan counties, reduced employment in the nationalized industries and public sector (down from 29.4 per cent of the workforce in 1979 to 26.9 per cent in 1985) and a lower number of trade unionists and council tenants, means that a future Labour government will start from a much weaker position in terms of promoting its traditional objectives of public ownership, trade union bargaining rights, and collectivism.

There is also some evidence that Mrs Thatcher has attracted a particular electoral constituency. That section of the middle class which includes shopkeepers, foremen, and the small and self-employed businessmen—the *petit bourgeois*—has turned strongly to the party since she became leader. By 1974 many of these people had become disillusioned with the Heath government. Yet a survey study of the 1983 election found that this section of the electorate was the most right-wing of all on issue questions about privatization, incentives versus equality, trade union rights, and comprehensive schools, and that over 70 per cent of them voted Conservative in the election.[28] The

Conservative electorate in the 1980s contains two significant groups. With 36 per cent of the manual workers' vote in 1987 the party's share of the working-class vote was the highest for any post-war election. It was higher still in what Crewe calls the 'new working class'—workers living in the south, owning their own homes, employed in the private sector and not in trade unions. The second group is concentrated among a particular section of the middle class—those in the private rather than the public sector, and among those who are not graduates.[29]

Some of the new thinking may already be common ground. The fine print of the speeches of Labour leaders and party policy documents has changed during the 1980s. The party now accepts Britain's membership of the European Community and in the 1989 Euro-elections claimed to be the party of 'good Europeans'. It has turned away from old style nationalization and is more willing to admit the need for enterprise and a role for markets. The party leadership has also accepted the case for state aid for regular elections of trade union officials by members, pre-strike ballots, and the sale of council houses to tenants. The 1989 policy review has made it clear that complete renationalization of privatized industries will not be a high priority for a future Labour government, that direct taxation will not be increased for those on average and slightly above average incomes, and that it is 'concerned' about law and order. It has also abandoned unilateralism and refuses to promise that a Labour government will rid Britain of nuclear weapons in the lifetime of the government. Compared to 1983, let alone 1979, Labour appears to be more aware of a changed electorate, one which is not largely poor, unemployed, in trade unions, working in heavy manufacturing, or living in council houses. A test of a shift in the central ground of politics is the extent to which the opposite party accepts, however reluctantly and implicitly, the new ideas. On this test, Thatcherism has achieved a partial success.

A by-product of Labour leaders' responsiveness to the new thinking may be that the party relieves itself of some commitments which have been electoral liabilities in the past. So far the leadership's reaction to the 1983 and 1987 defeats has differed from that to such earlier decisive election defeats in 1931, 1951, 1970, and 1979. Those defeats unleashed bitter internal conflicts and, except for 1951, a shift to the left on policy. The parliamentary leadership was usually assailed by the left for having failed to carry out socialist policies and having compromised with capitalism when in government. Since 1983 the leadership, having read the election defeats as a repudiation of left-wing policies, has been more concerned to reassure the electorate and attract the millions of 'missing' Labour voters. Militant Tendency in Liverpool and elsewhere, Mr Scargill of the NUM and Tony Benn have been lashed by the leadership as the unacceptable face of Labour and all have been marginalized in the party.

Among more radical monetarists and free marketeers there remains disappointment that the government had not done more to shift the middle ground. Such critics want bolder measures to cut taxation and state spending. In education radicals welcome the introduction of the assisted places scheme, CTCs, proposals for opting out, and increase in places on school governing boards for parents, but seek further reforms to erode the dominance of the state and extend parental choice. A scheme of school vouchers was considered and abandoned although a scheme for loans in higher education is to be introduced. No action has been taken to trim tax relief for pension contributions or house mortgages (although the Treasury has ended the entitlement to tax relief for two mortgages on one property and has maintained a limit of £30,000 on mortgages entitled to tax relief). In December 1985 the long-awaited Fowler review of the State Earnings Related Pension Scheme (SERPS) produced only a reduction in the pension benefits rather than an abolition of the scheme (it is worth noting, however, that the encouragement to leave SERPS for private schemes has had some success; by mid-1989 over one million people had left). On the NHS, for all the managerial changes and efficiency campaigns, Mrs Thatcher has made clear that it will remain a largely state-financed service.

Finally, the much-heralded Medium Term Financial Strategy (see p. 228) and monetarism were cast aside. In October 1984 Nigel Lawson's Mansion House speech formally abandoned a specific range for the M3 measure of the money supply. The Chancellor made clear that henceforth he would rely more on the narrower definition of money, M0, and maintain a high exchange rate (and high interest rates) to dampen inflation. In fact money supply grew rapidly and by 1989 inflation was over 8 per cent.

Some right-wing critics of the Conservative government have argued that the partial success in changing attitudes was reflected in the failure of government to be more radical in cutting the levels of taxation and public spending and the size of the public sector. Here is a critique of the Thatcher record from a different perspective than that of the left or of the centre. An editorial in *The Times*, a firm supporter of Thatcherite policies, could ask on 3 May 1985: 'If these obstructions to a society, based on enterprise and shorn of its collectivist illusions, cannot be dismantled in six years, what evidence is there that twelve years will be any better?' The paper claimed that the government had failed to make more progress in changing attitudes because so many beneficiaries of the tax state were opinion-formers and hostile to the government's goals and to free enterprise. The beneficiaries included teachers in higher education, bureaucrats, trade union leaders, welfare organizers, and leaders of public corporations. Critics pointed to the government's fear of losing votes among middle-class interests and to the power of particular lobbies.

According to a disappointed editorial in *The Economist* (15 June 1985) history would judge the Thatcher government as only 'an averagely wet Tory administration'. *The Times* again complained about the government's loss of radical purpose. In an editorial on 2 November 1985, headlined 'Unfinished Business', it asked 'whether the structures and institutions of British life have been so shaken by six years of Mrs Thatcher that they can no longer fall back into their old ways—or whether more shaking is needed?' It answered: 'More shaking is still needed'. This view was shared by most of the lobbies and opinion-forming groups discussed in Chapters 3 and 4. Their view was that the government lacked a clear strategy in its second term. By May 1989, however, many of these critics had come round to praise the continued radicalism of the Thatcher government, in education, tax cuts, and further measures of privatization.

An International Shift

Many of the trends in British politics in the 1980s have had counterparts in other western states. From the mid-1970s the slow-down in economic growth and rising inflation sapped the confidence of policy-makers in demand management. Strict control of the money supply seemed to be linked to low rates of inflation in Switzerland and West Germany. As many western states simultaneously suffered sharp increases in levels of unemployment and prices, concern grew over the growing costs of government welfare programmes, many of which were linked by statute to changes in price levels. The overall cost of welfare programmes in Western Europe was rising by an average of 7 per cent per annum in the 1970s, while the economy was growing by an average of 3.5 per cent.[30] Double-digit inflation and the growth in the costs of public policy at a rate faster than that of the economy posed obvious problems. The success in California of Proposition 13 in 1978 was paralleled by similar tax backlash movements in Denmark and the swing to the political right in Britain, Sweden, and Norway.

Not surprisingly the above trends have raised greater electoral and ideological difficulties for the political parties of the left. Parties associated with policies of increasing taxation and growing public expenditure appeared to be less popular at the ballot box. Many of the political values and supporters (particularly public-sector workers and welfare lobbies) of left-wing parties have favoured public spending and looked to governments as problem-solvers. However policy-makers' overriding concern with inflation and the calculations about likely reactions of financial markets to inflation-risking economic policies (notably in Britain in 1976, France in 1982–3, and Australia in 1984) have forced advocates of state spending on to the defensive. Policies of economic redistribution to the less well off met with resistance from skilled workers at a time of low economic growth.[31] A problem for many parties of the left in the era of slow growth was to reconcile high public spending and greater social and economic equality with the reluctance of many workers to pay heavier taxes. In Britain, West Germany, and Scandinavia in the 1970s there was a steady reduction in working-class support for parties of the left.

Politically there has been a shift to, or confirmation of, right-dominated governments for much of the 1980s in West Germany, Britain, USA, Denmark, the Netherlands, Belgium, Canada, and Japan. In the United States the long dominant Democratic New Deal electoral coalition, which Franklin Roosevelt created out of the white south and the industrial working-class, has been in retreat. The Democratic party has won the presidency only once out of the last six elections since 1964. The nomination of weak candidates has been a factor. But the party has also lost out because of the salience of race (which encouraged the white south to defect to the Republicans), embourgeoisement of the working class, and the Social Issue (concern about the rise of the anti-Vietnam war protest movement in the 1970s and the assertiveness and challenges to traditional values by various minority groups). President Reagan, like Margaret Thatcher, was a product of the concern in the 1970s over the economy, big spending programmes, social indiscipline, and, for the USA, military decline. It was Reagan, but it could have been Thatcher, who said in 1981, 'Government is not the solution to our problems. Government is the problem.' The litany of tax reductions, spending cuts, high defence spending, and, until the advent of Mr Gorbachev, anti-Soviet rhetoric, dominated Reagan's (and Thatcher's) administration.

In Denmark the normal pattern of Social Democratic government was disturbed by the emergence in 1982 of a four-party centre-right coalition, headed by a Conservative Prime Minister—the first since 1905. In Norway the multi-party system was further fragmented by the 1972 referendum on Norway's membership of the European Community. Since 1973 there has been a steady increase in electoral support for the Conservative party which dominated centre-right coalition governments in the early 1980s. In Sweden the Social Democrats were ousted from power in 1976 after forty-five years in office by an alliance of 'bourgeois' parties, although they returned to office in 1982. Yet the ideas of deregulation, limits to taxation, and budget-cutting have been accepted even by governments of the left in France, Spain, Australia, and New Zealand. Perhaps the most dramatic reactions against central state planning and a willingness to experiment with market forces have been seen in China, the USSR, Poland, and Hungary. By the end of the 1980s central economic planning was in retreat, even in its heartland.

It is possible to discern some international trends in economic policy. One has been a pursuit of 'prudent' finance and curbing inflation. The new package includes cutting income taxes, particularly the high marginal rates, containing public spending, 'privatizing' state assets (although Britain has gone further than any other state in this) and reducing loss-making activities among state enterprises. In Austria there have been sales by the holding company for state companies, Obesterreichissche Industrien, and in Sweden also. The socialist government in Spain has partly privatized state industries, including Iberia Airlines. In France there has been partial privatization of banking, industrial, and insurance groups. Another plank has been the encouragement of the private sector by deregulation (particularly in Britain and the United States), especially in transport, tax deductions for investment, and packages of incentives for small business. As in Britain this qualified support for the free market has been accompanied by measures for a stronger stance in defence and law and order (which have increased their shares in public spending in most Western European states).

A final trend has been the attempt to contain the growth in welfare and social security spending. Social security accounts for about a third of total public spending in most western states. It has been an obvious target for economy-minded governments. For example, Norway imposed a freeze on social security benefits and West Germany made all student grants repayable to the government, increased contributions to pension and sickness benefit, and delayed rises in pensions. Denmark has made cuts in employment and sickness benefits and France required hospital patients to contribute towards non-medical services and uprated welfare benefits in line with current inflation rather than the higher level of the previous year. Governments everywhere have tried to educate voters about the dilemma of the tax burden and state spending.

In spite of such cuts total social security spending continued to increase as a proportion of GDP—by an average of 5 per cent in real terms—in OECD states between 1975 and 1985. For all the talk of a tax backlash and attempts to cut state spending on welfare the irony is that both are now higher as shares of GDP in all OECD states compared with 1975 or 1980. Some of the rise has been prompted by the increased unemployment and rising pension and health demands of the growing army of the over-65s. Perhaps the main achievement of the political right has been to slow down the rate of growth in public spending and to identify some of it as 'wasteful'. Tax cutting may also have passed its peak. In Norway and Denmark electoral support for anti-tax parties has declined, while opinion surveys in the United States and Britain show that tax-cutting has become a minority cause, compared to support for spending on social programmes. Since 1978 most tax-cutting referendums in the United States have been defeated. There has been a general shift to less visible forms of taxation such as payroll (or national insurance) and indirect taxes.

As in the case of Britain one has to tread carefully in arguing that public opinion has moved to the right in other states in the past decade. In spite of the overwhelming popularity of Ronald Reagan, surveys show that the American public consistently refused to support many of his programmes. On social and moral issues, such as school prayers, the Equal Rights Amendment (for women), and abortion, the 'hard' positions of Reagan and the Moral Majority were in a minority. The majority of voters also favoured cuts in the defence budget and level or increased spending on domestic education and health programmes. Americans liked Reagan's personality and his handling of the economy and foreign policy, but they did not endorse Reaganism.[32] In Scandinavia the shift to the right has been more complex. In Norway and Denmark there continue to be high levels of popular support for state-provided welfare but there has been a marked and sustained shift against state control and intervention in the economy. In Sweden, however, approval for the welfare and economic interventionist aspects of collectivism has remained high.[33] The broad picture is not dissimilar from the British reaction to Thatcherism. In spite of the shift to the political right there remains widespread support for state-funded programmes in education and health and for those which are targeted at the poor and the old. In other words, there is extensive support in Britain and many other western states for government welfare and economic programmes which help the 'deserving' poor, and are also thought to be credible and realistic. Opinion surveys show attitudes that run across the old left–right divide. There is approval of the free market for wealth creation and many of the economic policies this entails. But there is also widespread support for state or collective provision of welfare even over tax cuts. The right-wing reaction has been half successful, repudiating Keynes but not Beveridge.

It is difficult to assess whether these trends will produce a lasting shift in the middle ground. Will they develop incrementally along existing lines or will there be a change cyclically to something resembling the old consensus? At present it appears that advocates of big spending programmes, expanding the public sector, inflation-risking policies, and increasing state control over the economy are in a minority. The vocabulary of many young leaders of left-wing parties in Western Europe—Kinnock, Gonzales, Rocard, Rau—refers to a 'servant' or 'opportunity' state, dismisses throwing money at problems as a solution, and promises to provide incentives for free enterprise. In debates over the Labour party's policy review in 1989, Mr Kinnock countered complaints from his left wing that he was propping up capitalism with the claim that Labour's task was to make the market work more efficiently and justly. So much for Clause IV. The 'new realism' of the left is more willing to acknowledge the limits of public spending and the need to encourage entrepreneurship.

A Changing Party System

The 1983 and 1987 general elections were interesting for psephological as well as for political reasons. Despite presiding over large-scale unemployment the Conservatives were returned to office by huge majorities. Although it is already fashionable to explain away the scale of the election successes—by such factors as the Falklands, Foot, Benn, and an inept Labour campaign in 1983, and prosperity and the disproportional electoral system in 1987—the results marked a further stage in the decline of the Labour party. By the next general election (say 1991) the Conservatives will have been in office for sixteen of the twenty-one years since 1970. After the 1987 election there was speculation that Britain had moved into an era of one-party politics or a predominantly one-party system. The Conservatives have been a long way ahead of Labour not just in seats (an arbitrary effect of the electoral system) but also in votes. Between 1950 and 1979 neither Labour nor Conservative ever had as much as 55 per cent of the combined two-party vote. But in the 1983 and 1987 elections the Conservatives averaged nearly 60 per cent and led Labour by 14.8 per cent and 11.5 per cent of the votes respectively in the two elections. Opposition in Parliament in the 1980s came more effectively from Conservative dissidents and the House of Lords than the Labour party. Not since the 1930s has the party system been so one-sided in Parliament.

One must always be wary of extrapolating from present trends to the future in electoral matters, as in others. The 1959 election defeat and the growing embourgeoisement of the working class were thought to spell future electoral ruin for Labour. The conditional projections made in 1969 by David Butler and Donald Stokes about the differential fertility—death rates between the social classes, combined with the transmission of political loyalties from parents to children—appeared to make Labour a natural majority party.[34] The first projection was immediately disconfirmed by Labour winning the 1964 and 1966 elections and the second was undone by the 1970 election and has been destroyed by the last three general elections. A telling statistic in a recent election study by Rose and McAllister is that, whereas in 1964 80 per cent of those who came from homes in which both parents were Labour voters themselves voted for the party, by 1983 the proportion from such backgrounds which still voted Labour had dropped to 45 per cent.[35]

TABLE 10.4 *The Fragmented Electorate 1923, 1929, 1983, 1987* (per cent)

	1923	1929	1983	1987
Conservative	38.1	38.2	42.4	42.3
Labour	30.5	37.1	27.6	30.8
Liberal/Alliance	29.6	23.4	25.4	22.6

The loss of working-class support for Labour is the outstanding feature of post-war electoral sociology in Britain. The 'new' affluent working class was clearly Conservative in 1983, and was so again in 1987. Ivor Crewe has drawn attention to the political significance of the more important social trends: twice as many manual workers are now in the private sector as in the public sector; there are far more home-owners than council tenants among them, and nearly as many live in the south as in the north and Scotland combined. Moreover, these trends are likely to increase (aided by the Conservative government) as a result of privatization, council house sales, and the growth of home-ownership and the southward drift of population. The 'traditional working class' is Britain's newest minority and it is far too small by itself to elect a Labour government.[36]

Few general elections or administrations map out a new agenda. Yet such a change can happen. A different set of policies, perhaps sustained over two or three elections, may force the defeated party to reconsider its strategy if it wants to remain electorally competitive. The adaptation of the Conservative party to the post-war Labour party initiatives is one example. The acceptance of the New Deal by many Republicans in the United States and the adoption of the social market approach and formal abandonment of Marxism by West Germany's Social Democrats are others. In the last two cases, decisive and successive election defeats convinced the party leaders that the centre of electoral opinion had moved away for them. Parties adopt or disavow policies not only to win forthcoming elections but also as a response to past electoral outcomes.

An election may also be a realigning one in that it produces a long-term shift in the balance of support between the parties, perhaps installing one as the new majority. The 1983 election, although a significant Conservative success, was not a realigning one, nor one necessarily presaging a long spell of Conservative rule. Prosperity was an important factor for many Conservative voters in 1983 and 1987; such votes are instrumental and conditional. Public opinion, though skewed to the right on many issues, has not moved further right between 1979 and 1989, and may actually have moved in a contrary direction. The proportion of voters identifying with the Conservative party is about the same as in 1964 (although Labour's fall has been steep) and the party's share of the vote in 1983 and 1987 was actually slightly less than in 1979. It profited handsomely from the division of the non-Conservative vote between Labour and Alliance, and from the exaggerating effects of the British electoral system; compared to 1979, a decline of 1.5 per cent in share of the vote had produced a trebling of the government's overall majority in seats from 44 to 144 in 1983 and to 100 in 1987.

The 1980s has been a decade of three-party politics. The emergence of the Alliance posed the most formidable third-party challenge to the Labour/Conservative duopoly since the multi-partyism of the 1920s. Indeed the distribution of votes across the three parties in 1983 bears a striking resemblance to the elections of 1923 and of 1929 (Table 10.4). Talk of a realignment on the centre-left of the spectrum has a long history. It seemed more credible in the 1980s with the emergence of the Alliance and Labour's electoral collapse. The widespread agreement across the electorate on many issues increases the possibilities for voters to shop around the political parties, as well as for the political parties to attract voters who are not traditional supporters.

The electoral system has been a barrier to realignment. Labour in the north of Britain and the Conservatives in the south have so many safe seats that they can suffer a loss of popular votes without this being translated into an equivalent loss of seats. The proportions differed only slightly in the 1987 general election. In 1983 the Conservatives gained 85 per cent of the seats in Southern England for 50 per cent of the votes, Labour 55 per cent of the seats in Northern England for 37 per cent of the vote. The Alliance had no such electoral stronghold. Surveys showed that its support, like that for the Liberals, was diffuse but shallow and lacked distinctive issue bases or social constituencies, apart from a following in the well-educated professional groups.

Mr Kinnock, however, has moved Labour back to the centre and since 1987 the two former Alliance parties have been in disarray. For all the turbulence and excitement that erupted with the Gang of Four's break with Labour in 1981, Britain looked as much a two-party system at the end of the decade as it did in 1980. The Alliance route to a realignment of the party system is dead. The great opportunity for Labour is that it will again be the main repository of anti-Conservative votes.

The significant electoral battleground in the future is likely to be the middle class or non-manual working class, simply because of its size. Even if at the next general election Labour does better than ever before in attracting working-class support, say 80 per cent of the working class, it still would not have a majority of the vote. Labour's problem is that its core constituency in the manual working class has steadily declined over the years. The growth of private home-ownership and car-ownership, reduction in the number of voters dependent on council housing, expansion of self-employment (up from 1.9 million to 3 million between 1979 and 1989) and of employment in the private sector, and a fall in the public-sector employment, may also work against the party. The Conservative government has not of course been passive in the process of restructuring the social bases of electoral choice. Between 1979 and 1989 the reduction in public-sector employment, extension of private home-ownership (up from 55 per cent to 65 per cent) and share-ownership (up from 7 per cent to 22 per cent), increasing scope for private provision of services, and run-down of trade union membership (down from 54 per cent to 46 per cent of the workforce) have all been government objectives.

None of these trends in social and economic structure and party identification mean that Labour can never win again, nor that the Conservatives are sure to win. Research suggests that many manual workers who have bought shares in the privatized firms and industries or purchased council houses were already Conservative voters.[37] But in so far as such voters may have their partisanship reinforced the trends do present a major hurdle to the election of a Labour government.

Comparison with the fortunes of the left in Western Europe are encouraging for Labour, but also chastening. In the recessionary 1970s, most governments were turned out at elections or saw their share of electoral support fall. The misery index (the combined percentages of unemployment and inflation) reached new heights. Although there has been some shift in the 1980s to the political right in the United States and West Germany, the socialist share of the vote has hardly changed over the two decades in Austria, Scandinavia, West Germany, and Italy. It has increased substantially in France and Spain, and Labour governments have been re-elected in New Zealand and Australia. Only in Britain has the fall in popular support for the left been substantial. Indeed the scale of decline for such a major party is virtually unparalleled in any other state in post-war Western Europe.

The good news for Labour from a comparative survey is that such features as wider home ownership, affluence, and the *embourgeoisement* of the working class are not necessarily electorally adverse. Political behaviour is not shaped exclusively by social forces. Democratic socialist parties can still thrive in prosperous societies as the examples of Sweden and France prove. It is worth adding that, compared to some other socialist parties, Labour has not been handicapped by the opposition or suspicion of a powerful Catholic Church (as in some Western European states), or divided by a powerful Communist movement in party politics and in trade unions (as in France and Italy). Moreover, because of the first-past-the-post electoral system and the predominantly two-party system, it has been one of the few parties of the left able to form a government on its own. Elsewhere socialist parties have had to compete in multiparty and proportional electoral systems and usually share power in government. The question that such analysis then suggests is: why has Labour done so badly? Perhaps other parties, simply because they have lacked the British Labour party's advantages, have necessarily had to be more adaptable and willing to make alliances. Left governments in Australia, New Zealand, France, and Spain have pursued a series of economic and social policies that are not dissimilar from those that Mrs Thatcher's government has pursued since 1979.

Conclusion

The rhetoric of party politics, particularly at election time, is often adversarial and absolutist in tone. The introduction of new men and new women promises a new beginning. Edward Heath, in his first speech to a Conservative party conference as Prime Minister in 1970, claimed: 'We were returned to office to change the course and the history of this nation—nothing less.' In each election manifesto between 1970 and 1983, Labour promised a 'fundamental and irreversible shift in the balance of power and wealth to working people'. In a foreword to the 1979 Conservative election manifesto, Mrs Thatcher suggested that the election might be the last chance voters had to reverse the extension of state power at the expense of the individual. In 1983 she spoke of the general election providing a 'choice' between two ways of life; before the 1987 election she spoke of the election as the opportunity to 'kill off' left-wing socialism.

The controversy surrounding the overall record of the Thatcher administration suggests that it has had a major impact. On the political left, the government is regularly reviled for running down the welfare state, smashing the trade unions, and waging class war. 'The most reactionary government this century' is merely a mild form of left-wing rhetoric. Some of Mrs Thatcher's 'cheerleaders' in the party and popular press see her as having achieved a virtual revolution in British politics, restoring the authority of government, putting the trade unions in their place, taming a greedy and parasitic public sector, and regaining a greater freedom of choice for people in many areas.

The case of the 1979–83 administration seems to support Richard Rose's claims about the weakness of party government. The gloomy trends of macro-economic indicators of economic growth, inflation, unemployment, and public spending over the period were in line with those for earlier governments. Such factors, however, are only partly under the control of government. The significant discontinuities elsewhere support the thesis of adversary politics.[38] These include the reduction in the legal immunities and rights of trade unions, rejection of formal incomes policy, and the tripartite style of decision-making, according priority to the abatement of inflation, even with unemployment at over three million, privatization of state industries and services, changes in the framework in which many public services are carried out, interventions in local government structure and imposition of far-reaching controls over its finance, changes in the welfare state, and open hostility to the civil service and large parts of the public sector. The government has challenged many of the assumptions of several participants and commentators on British politics.

The more one looks back on the record the more one is impressed by the role of Mrs Thatcher as a dominant figure and as an illustration of the power of the Prime Minister. Chapter 3 explored the various critiques of social democratic values and policies and the suggested policy alternatives. The critiques and the alternative policies found an audience because of a number of worrying developments in the mid-1970s (Chapter 5), independently of the campaigns of Sir Keith Joseph and Mrs Thatcher. We are not claiming that personal leadership is all important but rather that Mrs Thatcher's personality and policies enabled her to take advantage of the constellation of events and ideas. This is the more remarkable in view of the lukewarm support some of the policies have had in the Cabinet and party. Over time she has retained her ideological commitment, and the shortcomings as well as the successes of her government record are invoked by her as reasons to press on with the original strategy. She has clearly set a lead in a number of areas, pushing radical ideas and reforms (see Chapter 9). Although the discontinuity in economic policy from the Heath Cabinet (1972–4) is remarkable, there has been only one dissenting resignation from her Cabinet, partly on grounds of her style as well as policy (that of Mr Heseltine over Westland in January 1986).

Previous great reforming administrations—notably the Liberals (1906) and Labour (1945)—did not retain their parliamentary majorities for long. Both had been returned by huge majorities, having long been out of office. Both rapidly alienated many groups and the Liberal majority was extinguished by 1910 and Labour's by 1951. Most British governments have been worn down by failures (particularly economic), scandals, lack of purpose, and electoral boredom. Mrs Thatcher's governments have not only won elections but retained their drive. Mrs Thatcher still tells interviewers and staff that 'There is much more to do'. If there is a time to sew and a time to reap, the latter has not yet arrived.

But radical governments depend in large measure on opportunities—for example, an inept political opposition or a crisis —and have to catch a mood. In due course, they will suffer from the mood of 'time for a change', either to consolidate or reverse policies, or from the rise of new issues. The impact of the

Thatcher years can only be fully assessed in the future. What is clear, however, is that she dominated the content and style of British politics in the 1980s and shaped the politics of the post-Thatcher era.

Notes

1. P. Riddell, *The Thatcher Government* (Oxford, Martin Robertson, 1983), p. 38.
2. R. Rose, 'Parties, Factions and Tendencies in Britain', *Political Studies* (1964).
3. For a good discussion of these issues see Samuel Brittan, *A Restatement of Economic Liberalism* (London, Macmillan, 1989).
4. (London, Temple Smith, 1978).
5. The next paragraph draws on my essay 'Is Mrs Thatcher a Conservative?' in D. Kavanagh, *Personalities and Politics* (London, Macmillan, 1990).
6. *Inside Right* (London, Quartet, 1978), p. 121 and *Britain Can Work* (Oxford, Martin Robertson, 1983), p. 95.
7. Lord Whitelaw, *The Whitelaw Memoirs* (London, Aurum Press, 1989), p. 2.
8. J. Bulpitt, 'The Thatcher Statecraft', *Political Studies* (1986).
9. Ibid., p. 33.
10. Lord Hailsham, *Dilemma of Democracy* (London, Collins, 1978).
11. A. Gamble, 'Thatcherism and Conservative Politics', in S. Hall and M. Jacques (eds.), *The Politics of Thatcherism* (London, Penguin, 1983).
12. For statements on these lines, see D. Regan, *The Local Left* (Centre for Policy Studies, 1987) and N. Ridley, *The Local Right* (Centre for Policy Studies, 1989).
13. J. Gyford, *The Politics of Local Socialism* (London, Allen & Unwin, 1985), p. 71.
14. See *The Economist*, 27 May 1978, which contained details of a leaked party document about how a future Conservative government might cope with a coal strike. The strategy included new police tactics, building up coal stocks, recruiting non-union lorry drivers, and cutting off welfare benefits to strikers' families. These suggestions anticipated much of the government's strategy in the 1984–5 coal strike.
15. See A. Hartley, 'Beliefs, Politicians and Spiritual Times', *Encounter* (March 1985), pp. 22–6.
16. See D. Martin 'The Churches: Pink Bishops and the Iron Lady',

in D. Kavanagh and A. Seldon (eds.) *The Thatcher Effect* (Oxford, Clarendon Press, 1989).

17. For this kind of analysis see P. Nettl, 'Consensus or Élite Domination: The Case of Business', *Political Studies* (1967), R. Rose, 'The Variability of Party Government', *Political Studies* (1969), and J. Hoskyns, 'Conservatism is not Enough', *Political Quarterly* (1984).
18. 'Conservatism is not Enough'.
19. A report of the lecture is found in H. Young, 'The Mandarins Strike Back', *The Times*, 19 January 1984. For discussion of the relations between Sir Ian and Mrs Thatcher see P. Hennessy, *Whitehall* (London, Secker and Warburg, 1989), pp. 623–33.
20. On this see P. Hennessy, *The Great and the Good. An Enquiry into the British Establishment* (London, Policy Studies Institute. Research Report 654, 1986).
21. *The Thatcher Government*, p. 1.
22. For a presentation, and analysis of much survey data see I. Crewe, 'Values: the Crusade that Failed', in D. Kavanagh and A. Seldon, *The Thatcher Effect* and his 'Has the Electorate Become Thatcherite?' in R. Skidelsky (ed.), *Thatcherism* (London, Chatto and Windus, 1989).
23. A. Heath, R. Jowell, and J. Curtice, *How Britain Votes* (Oxford, Pergamon, 1985), pp. 136–7.
24. Ibid.
25. In R. Levitas (ed.), *The Ideology of the New Right* (Oxford, Polity Press, 1986), p. 52.
26. R. Rose, *Politics in England* (London, Faber, 1985), p. 293.
27. J. Alt, 'It May Be a Good Way to Run an Oil Company, but . . .: Oil and the Political Economy of Thatcherism', unpublished paper, read at Harvard University, April 1985.
28. A. Heath, R. Jowell, and J. Curtice, *How Britain Votes*, pp. 20–1.
29. See I. Crewe, 'Why Mrs Thatcher was Returned with a Landslide', *Social Studies Review* (1987).
30. R. Rose and G. Peters, *Can Government Go Bankrupt?* (London, Macmillan, 1978).
31. See J. Alt, *The Politics of Economic Decline* (Cambridge University Press, 1978).
32. For evidence see E. Ladd, 'On mandates, realignments and the 1984 presidential election', *Political Science Quarterly* (1985).
33. See J. Sainsbury, 'Scandinavian Party Politics Re-examined: Social Democracy in Decline?', *West European Politics* (1984).
34. *Political Change in Britain* (London, Macmillan, 1969).
35. R. Rose and I. McAllister, *Voters Begin to Choose* (London, Sage, 1986).

36. I. Crewe, 'The Electorate: Partisan Dealignment Ten Years On', *West European Politics* (1983), p. 195.
37. On this see P. Norris, 'Changing Social Cleavages and Left-Wing Support in Britain', unpublished paper, European Consortium for Political Research, Paris, April, 1989 and A. Heath, R. Jowell, J. Curtice, 'The Extension of Popular Capitalism', Political Studies Association, Warwick, April 1989.
38. For discussion of this debate see above, pp. 210–12.

SELECT BIBLIOGRAPHY

A MORE comprehensive guide to the relevant literature is available in the references contained at the end of each chapter. This brief bibliographical essay refers to materials which the author found particularly useful. I have written a lengthy essay on the substantial literature on Thatcherism, in 'Making Sense of Thatcher' in my *Personalities and Politics* (London, Macmillan, 1990).

INTRODUCTION (Chapter 1)

On the interaction between ideas and practice, S. Beer, *Modern British Politics* (London, Faber), first published in 1965, is still good value. W. H. Greenleaf's three volume *The British Political Tradition* (London, Methuen, 1983) is a classic study of political ideas in Britain and much else.

THE MAKING OF CONSENSUS POLITICS (Chapter 2)

On historical background, see A. Taylor, *English History 1914–45* (Oxford University Press, 1965) and K. Middlemas, *Politics in Industrial Society* (London, Deutsch, 1979). On the 1940s see P. Addison, *The Road to 1945* (London, Cape, 1975), H. Pelling, *The Labour Governments* (London, Macmillan, 1985), and K. Morgan, *Labour in Power 1945–51* (Oxford University Press, 1984). For 'What Went Wrong' see C. Barnett, *The Audit of War* (London, Macmillan, 1986). On the economy, S. Brittan, *Steering the Economy* (London, Penguin, 1971) is still good. The list of substantial biographies and memoirs for the main actors is now very large. J. Vaizey, *In Breach of Promise* (London, Weidenfeld and Nicholson, 1983), is useful. On the debate about consensus, see D. Kavanagh and P. Morris, *Consensus Politics from Attlee to Thatcher* (Oxford, Blackwell, 1989).

THE NEW RIGHT (Chapters 3 and 4)

Nicholas Bosanquet, *After the New Right* (London, Heinemann, 1983) was one of the first studies and remains useful on economic and social issues. Greenleaf's volume ii, is also helpful. The Institute of Economic Affairs has sponsored a number of volumes and one should read Arthur Seldon's edited *The Emerging Consensus* (London, IEA, 1981) and his *The New Right Enlightenment* (London, IEA, 1985), a set of libertarian essays. Sir Keith Joseph's *Reversing the Trend* (London, Barry Rose, 1975) and *Stranded in the Middle Ground* (London, 1976), the latter published by the Centre for Policy Studies, are essential. The left has been more concerned about the authoritarian features of the New Right. See R. Plant and K. Hoover, *The Rise of Conservative Corporatism* (London, Methuen, 1988) and Ruth Levitas, *The Ideology of the New Right* (London, Polity Press, 1983). For the United States see P. Steinfels, *The Neo-Conservatives* (Chicago, Simon and Schuster, 1979) and G. Peele, *Revival and Reaction* (Oxford University Press, 1984).

The Break-Up of Consensus (Chapter 5)

On general problems of 'governability', see A. King, 'Political Overload', *Political Studies* (1975) and S. Brittan, *The Economic Consequence of Democracy* (London, Temple Smith, 1977). Most of the literature deals with the political economy. M. Stewart, *The Jekyll and Hyde Years* (London, Dent, 1977), R. Skidelsky (ed.), *The End of the Keynesian Era* (London, Macmillan, 1979), R. Rose and G. Peters, *Can Government Go Bankrupt?* (London, Macmillan, 1978), S. Brittan and P. Lilley, *The Delusion of Incomes Policy* (London, Temple Smith, 1977), R. Bacon and W. Eltis, *Britain's Economic Problem: Too Few Producers* (London, Macmillan, 1976) are important. On industrial relations see M. Moran, *The Politics of Industrial Relations* (London, Macmillan, 1977) and C. Crouch, *The Politics of Industrial Relations* (London, Fontana, 1979). On 'What's wrong with' parties and politics, S. Finer (ed.), *Adversary Politics and Electoral Reform* (London, Wigram, 1975) is important. More generally, see D. Marquand, *The Unprincipled Society* (London, Cape, 1988).

Labour's Dilemma: Revisionism Revised (Chapter 6)

There is a large literature on the Labour party. In the last two decades it has increased immensely, come largely from the left of the political spectrum, and is often critical. These characteristics become more marked as the party is perceived to be in 'crisis'. A range of different concerns and stances is contained in L. Minkin, *The Labour Party Conference* (Manchester University Press, 1980). R. Miliband, *Parliamentary Socialism* (London, Allen & Unwin, 1967), H. Drucker, *Doctrine and Ethos in the Labour Party* (London, Allen & Unwin, 1979), D. Kavanagh (ed.), *The Politics of the Labour Party* (London, Macmillan, 1969), and D. Howell, *British Social Democracy* (London, Croom Helm, 1980).

A More Resolute Conservatism (Chapter 7)

R. Blake, *The Conservative Party from Disraeli to Thatcher* (London, Fontana, 1985), Z. Layton-Henry (ed.), *Conservative Party Politics* (London, Macmillan, 1980), and P. Norton and A. Aughey, *Conservatives and Conservatism* (London, Temple Smith, 1981) are good for background. A. Gamble is always interesting and his *The Conservative Nation* (London, Routledge and Kegan Paul, 1974) is no exception. There have been many analyses of the Thatcher government's economic record to date. See for example W. Keegan, *Mrs Thatcher's Economic Experiment* (London, Alan Lane, 1984), A. Walters, *Britain's Economic Renaissance* (Oxford University Press, 1985), G. Maynard, *The Economy under Mrs Thatcher* (Oxford, Blackwell, 1987) and N. Crafts, *British Economic Growth before and after 1979* (London, Centre for Economic Policy Research, 1989).

The Conservative Record 1979–1989 (Chapter 8)

P. Riddell, *The Thatcher Government* (Oxford, Martin Robertson, 1983 and 1987) is comprehensive and fair-minded. P. Jackson (ed.), *Implementing Government Policy Initiatives: The Thatcher Administration 1979–83* (London, RIPA, 1985) contains various authoritative studies of policy areas by scholars. M. Holmes, *The Thatcher Government 1979–1983* (London, Wheatsheaf, 1985) is a spirited but uncritical defence. R. Rose, *Do Parties Make a Difference?* (London, Macmillan, 2nd edn. 1984) is a counter to Finer. R. Skidelsky (ed.) *Thatcherism* (London, Chatto and Windus, 1989) is good in parts. From the left see A. Gamble, *The Strong State Free Economy* (London, Macmillan, 1988 and B. Jessop *et al.*, *Thatcherism* (Oxford Politics, 1988).

Mrs Thatcher: A Mobilizing Prime Minister (Chapter 9)

N. Wapshot and G. Brock, *Thatcher* (London, Fontana, 1983) is good for background. P. Hennessy, *Cabinet* (Oxford, Blackwell, 1986) and H. Young and A. Sloman, *The Thatcher Phenomenon* (London, BBC Publications, 1986) draw on views of people who have worked with and observed Mrs Thatcher. The autobiographies, J. Prior, *A Balance of Power* (London, Hamilton, 1986), F. Pym, *The Politics of Consent* (London, Constable, 1985) and Lord Whitelaw, *The Whitelaw Memoirs* (London, Aurum Press, 1989) do not add much. Young's massive biography, *One of Us* (London, Macmillan, 1989) is the best biography to date. A. King's essay 'Margaret Thatcher: The Style of a Prime Minister' in his *The British Prime Minister* (London, Macmillan, 2nd edn. 1985) is excellent.

THE REORDERING OF BRITISH POLITICS (Chapter 10)

J. Bulpitt, 'The Thatcher Statecraft', *Political Studies* (1986) and I. Gilmour, *Britain Can Work* (Oxford, Martin Robertson, 1983) are both stimulating. Gilmour sets out his idea of 'the traditional Tory approach to economics and politics'. Whereas he excludes Thatcher from this club, Bulpitt includes her. P. Jenkins, *Mrs Thatcher's Revolution* (London, Cape, 1987) is an assessment of the political changes. Contributors to D. Kavanagh and A. Seldon (eds.) *The Thatcher Effect* (Oxford, Oxford University Press, 1989) consider the effects in more than twenty years. On the psephology, see R. Rose and I. McAllister, *Voters Begin to Choose* (London, Sage, 1986), A. Heath *et al.*, *How Britain Votes* (Oxford, Pergamon, 1985), P. Dunleavy and C. Husbands, *British Democracy at the Crossroads* (London, Allen & Unwin, 1985), and D. Butler and D. Kavanagh, *The British General Election of 1987* (London, Macmillan, 1988).

INDEX